AF356557

Latest Advances in Electrothermal Models

Latest Advances in Electrothermal Models

Editors

Krzysztof Górecki
Paweł Górecki

MDPI • Basel • Beijing • Wuhan • Barcelona • Belgrade • Manchester • Tokyo • Cluj • Tianjin

Editors
Krzysztof Górecki
Gdynia Maritime University
Poland

Paweł Górecki
Gdynia Maritime University
Poland

Editorial Office
MDPI
St. Alban-Anlage 66
4052 Basel, Switzerland

This is a reprint of articles from the Special Issue published online in the open access journal *Energies* (ISSN 1996-1073) (available at: https://www.mdpi.com/journal/energies/special_issues/electrothermal_models).

For citation purposes, cite each article independently as indicated on the article page online and as indicated below:

LastName, A.A.; LastName, B.B.; LastName, C.C. Article Title. *Journal Name* **Year**, *Volume Number*, Page Range.

ISBN 978-3-0365-0334-9 (Hbk)
ISBN 978-3-0365-0335-6 (PDF)

Contents

About the Editors

Krzysztof Górecki received M.Sc. and Ph.D. degrees in electronics from the Technical University of Gdańsk, Gdańsk, Poland, in 1990 and 1999, respectively, and a D.Sc. degree from the Technical University of Łódź, Łódź, Poland, in 2008. He is the Dean of the Faculty of Electrical Engineering in Gdynia Maritime University, Gdynia, Poland, where he has been a Full Professor since 2016. His research interests are in the areas of modelling, and the analysis and measurement of semiconductor devices and electronic circuits, particularly those including thermal effects.

Paweł Górecki received M.Sc. and Ph.D degrees in electronics from Gdynia Maritime University, Gdynia, Poland, in 2016 and 2019, respectively. He is currently an Associate Professor in the Department of Marine Electronics in Gdynia Maritime University, Gdynia, Poland. His current research interests include modelling, analysis, and measurements of power semiconductor devices and power electronic networks containing IGBTs.

Preface to "Latest Advances in Electrothermal Models"

Dear Reader,

For many years, tendencies to miniaturize electronic systems and to increase the power density dissipated in components of these systems have been observed. During the operation of electronic components, their internal temperature increases due to self-heating phenomena and mutual thermal couplings between components located on the common base. An increase in the device's internal temperature causes changes in the characteristics of electronic components, and also causes a decrease in the lifetime of these components. Therefore, one of the biggest problems in electronics at present is the accurate calculation of the values of the internal temperature of such components. Solving this task requires accurate models of the considered components and circuits, which take into account all important physical phenomena occurring in these components and circuits.

Models including both electrical and thermal phenomena are called electrothermal models. Such models have been described in the literature for the last 50 years, but the development of electronic technology also leads to the development of electrothermal models of electronic components. New versions of such a class of models have different forms and are dedicated to different software for the computer analysis of electronic circuits, and they could also make it possible to shorten the duration of calculations. An effective electrothermal analysis also requires effective methods of estimation of the parameter values that exist in the used models.

In recent years, we can observe dynamic developments in the abovementioned electrothermal models. This Special Issue of *Energies* is devoted to the latest advances in this area and contains eight articles. The first article presents the results of electro-thermal simulations of SiC power MOSFETs using a SPICE-like simulation program. The manner of automatic generation of a compact thermal model from accurate 3D mesh is described. The second article concerns the problem of modelling the thermal properties of inductors. A thermal model of an inductor proposed in this article takes into account the influence of power dissipated in the core and in the winding of the inductor, and the volume of the core, on the efficiency of heat removal. The third article deals with the problem of inserting a temperature sensor in the neighbourhood of a chip to monitor the junction temperature. The fourth article proposes a method of computations of the internal temperature of power LEDs situated in modules containing multiple-power LEDs, taking into account both self-heating in each power LED and mutual thermal couplings between each diode. The influence of the thermal pad's surface area on the device temperature is investigated. The fifth paper proposes an electrothermal averaged model of the diode–transistor switch, including an IGBT and a rapid-switching diode. This model makes it possible to quickly calculate the values of currents, voltages and internal temperatures of the mentioned semiconductor devices contained in a dc-dc converter using DC analysis in SPICE. The sixth article proposes an electrothermal model of SiC power BJTs. It is proven that this model properly describes the DC electrothermal characteristics of the considered device. The seventh article contains an analysis of the efficiency of selected algorithms used to solve heat transfer problems at the nanoscale. The case study is an MEMS structure. The eighth article presents an analysis related to thermal simulation of the test structure dedicated to heat-diffusion investigation at the nanoscale. The results of computations and measurements are presented and discussed. We hope that you find this book interesting and useful.

Krzysztof Górecki, Paweł Górecki

Editors

Article

Circuit-Based Electrothermal Simulation of Multicellular SiC Power MOSFETs Using FANTASTIC

Vincenzo d'Alessandro [1,*]**, Lorenzo Codecasa** [2]**, Antonio Pio Catalano** [1] **and Ciro Scognamillo** [1]

[1]   Department of Electrical Engineering and Information Technology, University Federico II, 80125 Naples, Italy; antoniopio.catalano@unina.it (A.P.C.); ciro.scognamillo@unina.it (C.S.)

[2]   Department of Electronics, Information and Bioengineering, Politecnico di Milano, 20133 Milan, Italy; lorenzo.codecasa@polimi.it

*   Correspondence: vindales@unina.it

Received: 26 June 2020; Accepted: 30 August 2020; Published: 3 September 2020

Abstract: This paper discusses the benefits of an advanced highly-efficient approach to static and dynamic electrothermal simulations of multicellular silicon carbide (SiC) power MOSFETs. The strategy is based on a fully circuital representation of the device, which is discretized into an assigned number of individual cells, high enough to analyze temperature and current nonuniformities over the active area. The cells are described with subcircuits implementing a simple transistor model that accounts for the utmost influence of the traps at the SiC/SiO$_2$ interface. The power-temperature feedback is emulated with an equivalent network corresponding to a compact thermal model automatically generated by the FANTASTIC tool from an accurate 3D mesh of the component under test. The resulting macrocircuit can be solved by any SPICE-like simulation program with low computational burden and rare occurrence of convergence issues.

Keywords: electrothermal (ET) simulation; finite-element method (FEM); model-order reduction (MOR); multicellular power MOSFET; silicon carbide (SiC)

---

1. Introduction

Silicon carbide (SiC) power devices are promising candidates for energy distribution, as well as for automotive, aircraft, and spacecraft applications, by virtue of their inherent features like high breakdown voltage, low on-state resistance, and excellent high-temperature capability [1].

Unfortunately, such devices often operate under critical conditions with a large amount of heat generation, which may lead to reliability degradation or even to an irreversible device failure in harsh cases. As a consequence, reliable simulation tools accounting for electrothermal (ET) effects are highly desired to define the thermal dissipation constraints and optimize the design of the transistor layout and/or of the cooling system. Such tools must be suited to describe temperature and current nonuniformities, which are often responsible for the safe operating area shrinking of transistors with a multicellular pattern. However, conceiving and developing a viable simulation strategy are challenging tasks due to multiple reasons. Fully numerical 3D ET analyses with device simulators concurrently solving the semiconductor and heat transfer equations are computationally unfeasible. Commonly-adopted approaches rely (i) on the interaction between a circuit simulation program and a 3D thermal-only numerical solver in a relaxation procedure [2,3], or (ii) on the extension of a finite-volume-/-element software package to account for the electrical behavior of the transistor with simplified models [4,5]. However, for the specific case of SiC power devices, results can be trustworthy only by using models that accurately describe the key physical parameters and their *non-intuitive* temperature dependences, which are rather different compared to the traditional silicon (Si) counterparts. In addition, regardless of the technology, the pre-processing geometry/mesh construction within the environment of the thermal solver is onerous and troublesome. Lastly, these approaches are very resource-hungry and prone to convergence failures, especially if dynamic simulations under critical conditions have to be performed.

In this paper, an innovative *circuit-based* ET simulation approach is proposed for multicellular SiC power MOSFETs, with the ambition of optimizing the trade-off between computational efficiency and accuracy. The strategy is articulated as follows: (i) the device is discretized into an assigned number N of individual cells (each associated to an independent heat source) described with a simple, yet accurate model accounting for the relevant influence of SiC/SiO_2 interface traps; (ii) the cell model is implemented with a SPICE-compatible subcircuit; (iii) an exceptionally accurate 3D finite-element method (FEM) description of the device is effortlessly obtained through a commercial solver aided by an *in-house* routine; (iv) the FANTASTIC code [6] is invoked, which *automatically* generates a dynamic compact thermal model (DCTM) of the device from the FEM representation, and thus builds an electrical network emulating the power-temperature feedback; (v) such a network is enriched with suitable voltage sources to account for nonlinear thermal effects, and the resulting circuit is referred to as *thermal feedback block*; (vi) a *purely-electrical macrocircuit* describing the whole ET behavior of the power device is constructed by connecting the N cell subcircuits to the thermal feedback block; (vii) the macrocircuit can be solved by *any* commercial circuit simulator in very short times and with unlikely occurrence of convergence problems. A multicellular 4H-SiC power MOSFET soldered on a direct bonded copper (DBC) substrate and operated under dc, short-circuit (SC), and unclamped inductive switching (UIS) conditions is considered as a case study.

This work extends the preliminary contribution [7], where a fully circuital representation of the SiC power MOSFET was obtained and solved with a similar approach. However, the generation of the thermal feedback block, based on Foster networks, was carried out in a long pre-processing stage involving N 3D FEM transient simulations, as well as a procedure to determine the proper number of RC pairs of each network and identify their values; in addition, the thermal interactions among horizontally-far cells were either coarsely described or even disregarded.

The remainder of the paper is articulated as follows. In Section 2, the device selected to test the approach and the experimental setup are described. Section 3 offers details concerning the transistor model used for the elementary cells. Section 4 probes into the ET simulation approach. Results are reported and discussed in Section 5. Conclusions are then drawn in Section 6.

2. Device under Test and Experimental Setup

The whole analysis was performed on the CREE 4H-SiC vertical double-diffused MOSFET (VDMOS) denoted as CPMF-1200-S080B, rated 1200 V, 50 A, 80 mΩ, and targeted at solar inverters, high-voltage dc–dc converters, and motor drives. A top-view picture of the device under test (DUT) is given in Figure 1; indicated are the gate pad, the two source pads, and the gate interconnect tracks. The DUT presents a 4.08×4.08 mm^2 die area and a 3.46×3.46 mm^2 active area (there is a peripheral inactive region), the *effective* portion of which contributing to the current capability amounts to about 10 mm^2. The pattern is multicellular, with many thousands of body islands located in the N-drift region, within which there are source-body contacts surrounded by the polysilicon gate that lies beneath the source metal. The Ni/Ag drain contact is on the backside.

Isothermal measurements of *I–V* transfer and output characteristics of the bare die were performed by means of an *in-house* 250 A-rated curve tracer suited to apply down to 1 μs-wide current pulses, the device baseplate being set to assigned temperatures T_B through a thermochuck with 1 °C resolution. UIS experiments aimed at the evaluation of the breakdown voltage were carried out by a non-destructive custom tester [8]. The switching behavior was investigated by using a half-bridge converter board configured to perform a standard inductive load switching (ILS) test [9].

Figure 1. Device under test (bare die).

3. Transistor Model and Parameter Extraction Methodology

In the last decades, a noticeable effort has been made to develop models for SiC MOSFETs, a review of them being offered in [10]. Here we propose a slightly modified version of the behavioral model presented in [7,9] and used for ET simulations in [7,9,11]. Such a model, unlike those hitherto described in the literature, concurrently enjoys the following benefits: (i) it is simple, yet accurate enough, with a few parameters easy to extract; (ii) it can be implemented with a subcircuit compatible with *any* SPICE-like program; (iii) it includes all the key physical parameters and their specific temperature dependence up to very high temperatures; (iv) it also accounts for avalanche effects due to impact ionization (II); (v) the nonlinear nature of the intrinsic capacitances can be activated.

3.1. Transistor Model

The model is an enriched variant of the classic SPICE Level 1. In particular, the transistor is represented as the series of an "intrinsic" conventional MOSFET describing the channel behavior, and a resistance for the lowly-doped N-type epitaxial region, as depicted in Figure 2.

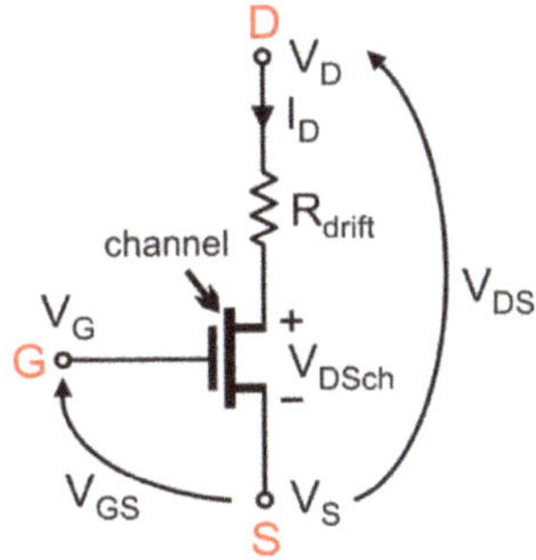

Figure 2. Sketch of the transistor representation.

Let us consider the following nomenclature:

- T [K] is the transistor temperature, assumed uniform in the regions impacting the device behavior.
- $T_0 = 300$ K is the reference temperature.
- $\Delta T = T - T_0$ [K] is the temperature rise over T_0.
- T_B [K] the baseplate (and thermochuck) temperature.
- V_{GS} [V] is the gate-source voltage.
- V_{DS} [V] is the drain-source voltage.
- I_D [A] is the drain current.
- R_{drift} [Ω] is the bias- and temperature-dependent resistance of the N-type epitaxial drift region.

- $V_{DSch} = V_{DS} - V_{drift}$ [V] is the V_{DS} portion falling over the channel ($V_{drift} = R_{drift} \cdot I_D$).
- V_{TH} [V] is the temperature-dependent threshold voltage.
- K [A/V^2] is the temperature-dependent current factor.

The channel region is described with the Level 1 model. If V_{DSch} is lower than the overdrive voltage $V_{GS} - V_{TH}$, the DUT operates in triode mode, and the II-free drain current I_{DnoII} is expressed as

$$I_{DnoII} = K \cdot \left[2 \cdot (V_{GS} - V_{TH}) \cdot V_{DSch} - V_{DSch}^2 \right] \tag{1}$$

Conversely, if $V_{DSch} \geq V_{GS} - V_{TH}$, then the DUT is driven into pinch-off, and

$$I_{DnoII} = K \cdot (V_{GS} - V_{TH})^2 \tag{2}$$

The negative temperature coefficient of V_{TH} is described through the following law:

$$V_{TH}(T) = \left[V_{TH}(T_0) - V_{TH\infty} \right] \cdot \exp\left(-a_{V_{TH}} \cdot \Delta T \right) + V_{TH\infty} \tag{3}$$

such an exponential model being an improvement with respect to the simple linear relationship used in [7].

The current factor K depends upon T since the electron mobility in the channel is temperature-sensitive; similar to [7,9], such a dependence is taken into account through the power relationship

$$K(T) = K(T_0) \cdot \left(\frac{T}{T_0} \right)^{-m(T)} \tag{4}$$

where the exponent $m(T)$ is

$$m(T) = -a_m + (a_m + b_m) \cdot \left[1 - c_m \cdot \exp\left(-d_m \cdot \frac{T}{T_0} \right) \right] \tag{5}$$

II effects are accounted for as follows [7,9]. The bias- and temperature-sensitive avalanche multiplication factor M (≥ 1) is given by [12]

$$M(V_{DS}, I_D, T) = 1 + m_{II} \cdot \tan\left\{ f_I(I_D) \cdot \frac{\pi}{2} \cdot \left[\frac{V_{DS} - R_{II} \cdot I_D}{BV_{DS}(T)} \right]^{n_{II}} \right\} \tag{6}$$

where $BV_{DS}(T)$ is the temperature-dependent drain-source breakdown voltage, expressed as

$$BV_{DS}(T) = BV_{DS}(T_0) \cdot \exp(\alpha_{II} \cdot \Delta T) \tag{7}$$

and $f_I(I_D)$ is a nondimensional correction term to describe a potential II dependence on current (i.e., on biasing conditions), given by

$$f_I(I_D) = \exp(\beta_{II} \cdot I_D) \tag{8}$$

Let us introduce the avalanche coefficient $\xi = M - 1$ (≥ 0). The II-affected drain current I_D is evaluated as

$$I_D = I_{DnoII} + I_{DII} = I_{DnoII} + \xi \cdot (I_{leak} + I_{DnoII}) \tag{9}$$

where I_{DII} is the additional current component only dictated by II, and I_{leak} is a small leakage current.

The resistance R_{drift} is expressed as the sum of (i) a bias- and temperature-dependent resistance R_{JFET} to model the path composed by the accumulation and JFET regions, and (ii) a temperature-dependent resistance R_{epi} for the epitaxial region beneath the JFET one [7,9]:

$$R_{drift}(V_{GS}, V_{drift}, T) = R_{JFET}(V_{GS}, V_{drift}, T) + R_{epi}(T) \tag{10}$$

where

$$R_{JFET}\left(V_{GS}, V_{drift}, T\right) = R_{JFET}(T_0) \cdot \left(\tfrac{T}{T_0}\right)^{m_{R_{JFET}}} \cdot \left(\frac{1}{1+V_1/V_{drift}}\right) \cdot \left(\frac{V_{GS}}{V_2}\right)^{-\eta}$$

$$R_{epi}(T) = R_{epi}(T_0) \cdot \left(\tfrac{T}{T_0}\right)^{m_{R_{epi}}} \tag{11}$$

$R_{JFET}(T_0)$ being the JFET resistance at $T = T_0$, $V_{drift} \gg V_1$, and $V_{GS} = V_2$ (V_1 and V_2 are fitting parameters). This formulation improves the one reported in [7] in the high-current triode region and is derived on the basis of simple arguments. First, the resistance of the accumulation region reduces with gate voltage due to the increased concentration of the attracted electrons; second, under high V_{drift} values, the high electric field occurring in the JFET region tends to saturate the electron velocity, thus degrading the mobility.

The dynamic transistor behavior is described by improving the Level 1 capacitance models; the nonlinear nature of C_{GD} and $C_{DS} = C_{DB}$ is accounted for with the following expressions [9]:

$$C_{GD}(V_{GD}) = (C_{GD0} - C_{GDMIN}) \cdot \left[1 + \frac{2}{\pi}\arctan\left(\frac{V_{GD}}{V^*}\right)\right] \tag{12}$$

and

$$C_{DS}(V_{DS}) = \frac{2}{\pi} \cdot C_{DS0} \cdot \left[\frac{\pi}{2} + \arctan\left(-\frac{V_{DS}}{V^{**}}\right)\right] + C_{DSMIN} \tag{13}$$

while C_{GS} was not modified [13].

3.2. Parameter Extraction Procedure

The threshold voltage V_{TH} and the current factor K were extracted in a very wide T_B range spanning from 300 to 500 K by resorting to the traditional *quadratic extrapolation method* [14,15] applied to I_D–V_{GS} transfer characteristics measured under isothermal (pulsed) conditions at $V_{DS} = 20$ V with the curve tracer mentioned in Section 2. Then the parameters in (3), (4), and (5) were calibrated so as to ensure the best matching between experimental data and the following relations:

$$V_{TH}(T_B) = [V_{TH}(T_0) - V_{TH\infty}] \cdot \exp\left[-a_{V_{TH}} \cdot (T_B - T_0)\right] + V_{TH\infty} \tag{14}$$

$$K(T_B) = K(T_0) \cdot \left(\frac{T_B}{T_0}\right)^{-m(T_B)} \tag{15}$$

with

$$m(T_B) = -a_m + (a_m + b_m) \cdot \left[1 - c_m \cdot \exp\left(-d_m \cdot \frac{T_B}{T_0}\right)\right] \tag{16}$$

The comparison between the measured V_{TH} and the optimized (14) is shown in Figure 3. It can be inferred that the DUT exhibits (i) a high $V_{TH}(T_0)$ (≈ 6.4 V) and (ii) a high negative temperature coefficient of $V_{TH}(T)$ at low/medium T_B compared to similarly-rated Si power MOSFETs. Both findings were attributed to the high density of SiC/SiO$_2$ interface traps (quantum states originating from the thermal oxidation of the SiC surface); more specifically, (i) electrons are captured by traps and do not contribute to channel formation, thereby leading to a high threshold voltage $V_{TH}(T_0)$, and (ii) V_{TH} markedly reduces with temperature due to the concurrent effect of more broken bounds that release electrons, and of the emission of inversion electrons from the traps, the latter effect being almost absent in the Si counterparts [16–18]. It must be underlined that the strong negative temperature coefficient in turn contributes to a significant positive temperature coefficient of I_D, which may exacerbate the ET feedback [19,20].

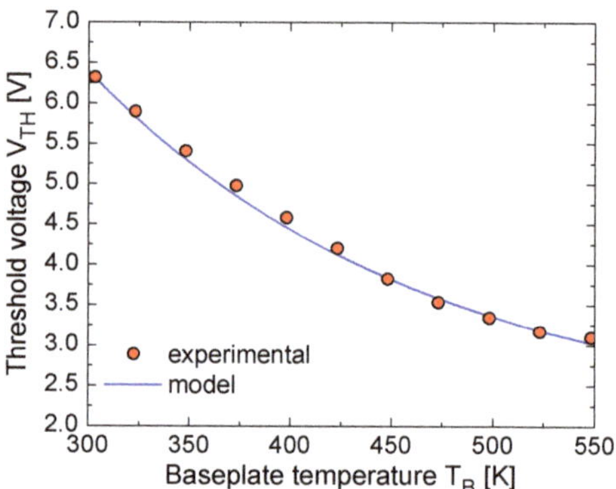

Figure 3. Threshold voltage V_{TH} vs. baseplate temperature T_B: comparison between experimental data (red circles) and model (14) (solid blue line) with optimized parameters.

The comparison between the measured K and model (15) and (16) with tuned parameters is shown in Figure 4. The temperature sensitivity of K is only related to that of the channel electron mobility μ_n, which is due to the interplay between (i) the Coulomb scattering with the filled (charged) interface traps, leading to a positive temperature coefficient induced by the trap discharging (release of electrons) with increasing temperature, and (ii) the acoustic-phonon scattering favoring a negative coefficient, where (i) prevails at low temperatures and (ii) dominates at high temperatures [18,21–23]. This behavior is accurately described by the m model (16).

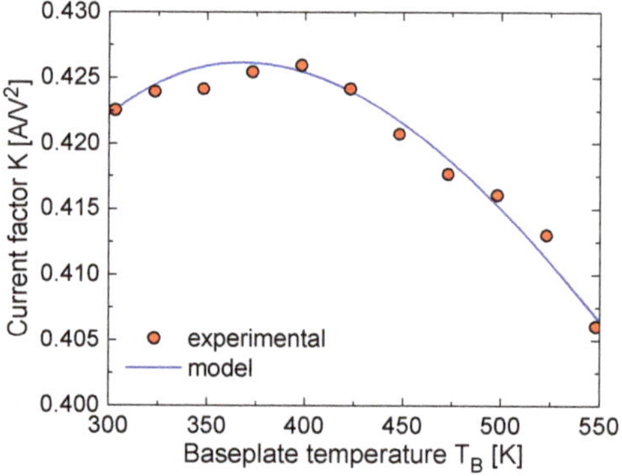

Figure 4. Current factor K against baseplate temperature T_B: comparison between experimental data (red circles) and model (15) and (16) (solid blue line) with optimized parameters.

The accuracy of the parameter calibration for the V_{TH} and K models is witnessed by the comparison reported in Figure 5 between (2) and the I_D–V_{GS} transfer characteristics measured under isothermal conditions at $V_{DS} = 20$ V and various T_B values in a current range wherein the DUT operates in pinch-off (note that $I_D \approx I_{DnoII}$). An inspection of the curves reveals the considerable positive temperature coefficient of I_D within a wide range of currents.

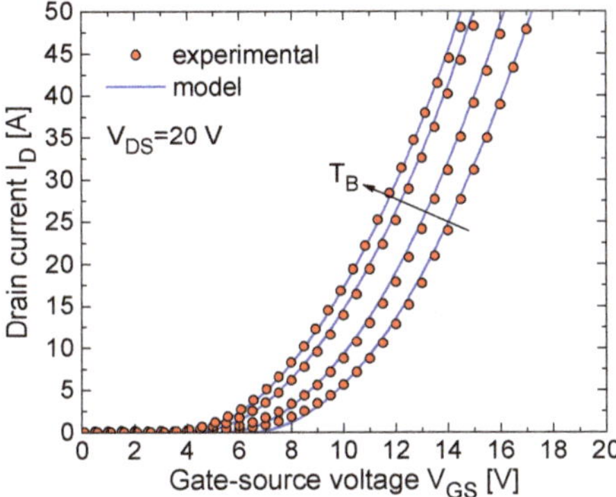

Figure 5. I_D–V_{GS} transfer characteristics of the DUT at V_{DS} = 20 V and T_B = 303, 348, 423, and 473 K: comparison between experimental data (red circles) and model (2) with calibrated parameters for the V_{TH} and K formulations (solid blue lines).

The parameters of the drift resistance R_{drift} in (10), (11) were optimized by comparing the I_D–V_{DS} output characteristics at various V_{GS} and different T_B values in triode region with the model (1) applied to the scheme in Figure 2. Figure 6 reveals the accuracy of the extraction at T_B = 303 K.

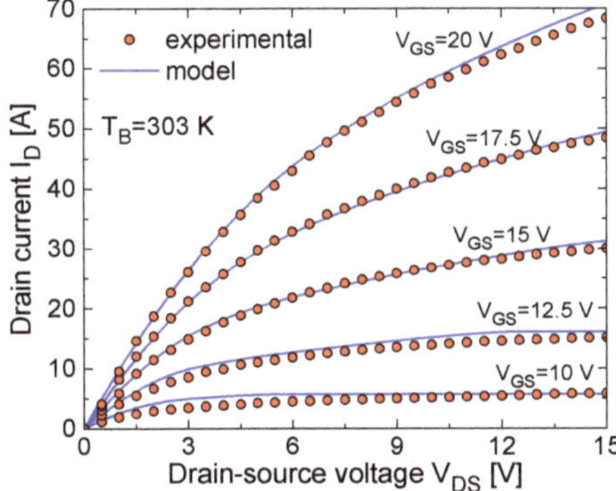

Figure 6. I_D–V_{DS} output characteristics of the DUT at V_{GS} = 10, 12.5, 15, 17.5, 20 V and T_B = 303 K: comparison between experimental data (red circles) and the model described in Section 2, where R_{drift} is given by (10), (11) with tuned parameters (solid blue lines).

The parameters of the II model given by (6) with (7) and (8) were tailored on the basis of experimental data determined under UIS conditions, and 2D numerical simulations of the T_B-dependent I_D–V_{DS} avalanche curve at V_{GS} = 0 V of an individual semi-cell of the DUT carried out with TCAD Sentaurus [24] (it must be remarked that the term "cell" here does *not* correspond to the discretization adopted in this work and identified by N, but it is one of the tens of thousands of identical blocks in which the effective active area of the MOSFET can be partitioned).

The parameters used in the expression of the capacitances C_{GD} (12) and C_{DS} (13) were calibrated to match the experimental gate and drain waveforms during an ILS turn-off and turn-on transient at different supply voltages.

All the optimized parameter values are reported in Table 1.

Table 1. Optimized parameter values used in the transistor model.

Parameter	Definition	Value
$V_{TH}(T_0)$	threshold voltage at reference temperature T_0	6.398 V
$V_{TH\infty}$	parameter of the threshold voltage model	2.05 V
a_{VTH}	temperature coefficient of the threshold voltage	6 mK^{-1}
$K(T_0)$	current factor at reference temperature T_0	0.422 A/V^2
a_m	parameter of the exponent m accounting for the mobility temperature dependence	0.24
b_m	parameter of the exponent m accounting for the mobility temperature dependence	2
c_m	parameter of the exponent m accounting for the mobility temperature dependence	1.02
d_m	parameter of the exponent m accounting for the mobility temperature dependence	0.09
$R_{JFET}(T_0)$	resistance of the accumulation and JFET regions at reference temperature T_0, $V_{drift} \gg V_1$, and $V_{GS} = V_2$	$0.235 \ \Omega$
m_{RJFET}	exponent for the temperature dependence of the resistance R_{JFET}	-1.3
V_1	parameter to account for the R_{JFET} dependence on V_{drift}	13 V
V_2	parameter to account for the R_{JFET} dependence on V_{GS}	20 V
η	exponent to account for the R_{JFET} dependence on V_{GS}	3.45
$R_{epi}(T_0)$	resistance of the deep epilayer region at reference temperature T_0	$10 \text{ m}\Omega$
m_{Repi}	exponent to account for the potential temperature dependence of the resistance R_{epi}	0
$BV_{DS}(T_0)$	drain-source breakdown voltage at reference temperature T_0	1750 V
m_{II}	parameter of the avalanche multiplication factor M	1.8
n_{II}	parameter of the avalanche multiplication factor M	2.9
α_{II}	temperature coefficient of the breakdown voltage BV	0.18 mK^{-1}
β_{II}	coefficient to account for the I_D dependence of factor M	0 A^{-1}
R_{II}	resistance to account for the I_D dependence of factor M	$10 \ \Omega$
C_{GD0}	zero-bias gate-drain capacitance	0.85 nF
C_{GDMIN}	minimum gate-drain reverse-biased capacitance	0.01 nF
V^*	gate-drain capacitance parameter	2 V
C_{DS0}	zero-bias drain-source capacitance	2.8 nF
C_{DSMIN}	minimum drain-source reverse-biased capacitance	0.06 nF
V^{**}	drain-source capacitance parameter	10 V

3.3. Impact of Interface Traps

The impact of the interface traps was investigated by resorting to the following strategy. The traps were virtually removed by (i) reducing $V_{TH}(T_0)$ to 4 V, (ii) decreasing a_{VTH} to 2 mK^{-1}, (iii) multiplying the current factor $K(T_0)$ by 50 to annihilate the degradation of mobility $\mu_n(T_0)$, and (iv) setting $c_m = 0$ to eliminate the influence of Coulomb scattering on μ_n, which will thus exhibit a negative temperature coefficient over the whole temperature range. Figure 7 shows the comparison between the *real* 4H-SiC DUT and the *ideal* traps-free counterpart in terms of transfer characteristics at $V_{DS} = 20$ V and various T_B values. It can be inferred that:

- differently from the DUT, which suffers from a marked positive temperature coefficient over the entire current range, the traps-free device is subject to a slight positive coefficient only at low drain current I_D (<20 A), where the negative temperature coefficient of V_{TH} prevails over the negative coefficient of μ_n, while beyond a zero-temperature coefficient region (also referred to as *compensation* region), the negative coefficient of μ_n dominates, and I_D reduces with temperature.
- the traps-free transistor benefits from a much higher current capability due to the lower V_{TH} and the higher μ_n.

To further corroborate the above findings, Figure 8 reports the temperature coefficient of the drain current I_D, given by [19,20,25]

$$\alpha_T = \left. \frac{\partial I_D}{\partial T} \right|_{V_{DS}} \tag{17}$$

for the real DUT and its traps-free variant at reference temperature T_0 and drain-source voltage $V_{DS} = 20$ V; it is again witnessed that the traps at the SiC/SiO$_2$ interface lead to a detrimental highly-positive α_T within a broad range of currents.

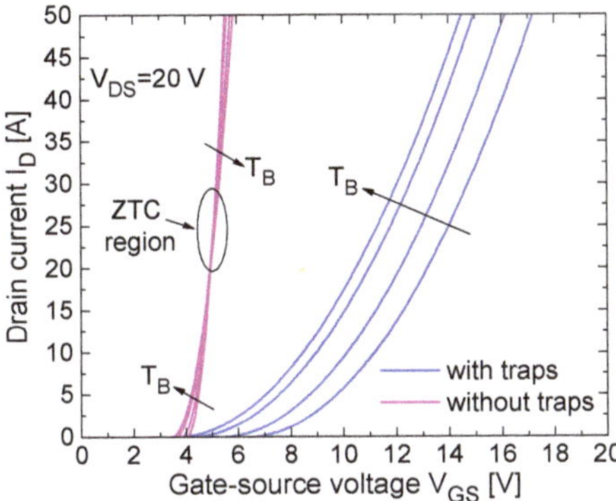

Figure 7. Modeled I_D–V_{GS} transfer characteristics at $V_{DS} = 20$ V and $T_B = 303, 348, 423,$ and 473 K: comparison between the real DUT (blue lines) and the traps-free counterpart (magenta). ZTC stands for zero-temperature coefficient.

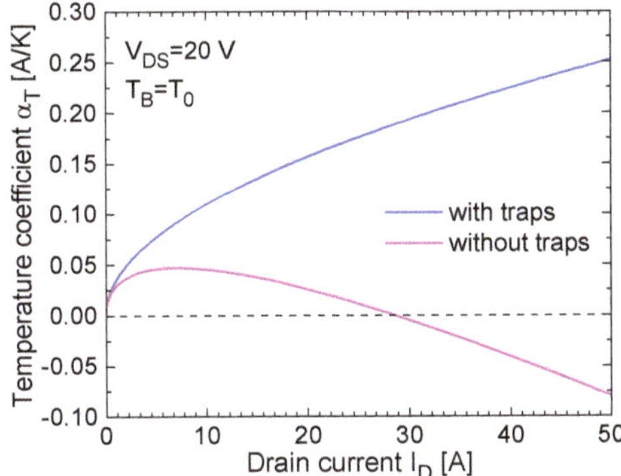

Figure 8. Temperature coefficient α_T of the drain current at $T_B = T_0$ and $V_{DS} = 20$ V for the real DUT (blue line) and its traps-free version (magenta).

4. Electrothermal Simulation Approach

We chose to resort to a *circuit-based* approach [7,11,26,27], which is a good trade-off between computational burden and accuracy. Such an approach makes use of the *thermal equivalent of the Ohm's law (TEOL)* and can be summarized as follows.

- The whole DUT is subdivided into an assigned number N of elementary cells, high enough to identify potentially-dangerous temperature gradients over the transistor active area, but not too high to prevent intolerably long CPU times or impossible memory storage. The individual cells are described with the model explained in Section 3, where the area-dependent parameters are properly scaled. For the simulation campaign reported in Section 5, all cells are assumed identical, although it is in principle possible to assign different parameter values to different cells to allow a statistical analysis of the influence of parameter fluctuations on the ET behavior of the component.
- Each cell is represented with a SPICE-compatible *subcircuit* implementing the above model through a *macromodeling technique*. The subcircuit makes use of (i) a standard MOSFET at reference temperature $T_0 = 300$ K as a "main" component to describe the channel region, and (ii) linear and nonlinear controlled sources to include all the model features that cannot be accounted for with the basic MOSFET, i.e., the temperature dependence of the threshold

voltage V_{TH} and of the current factor K, the bias- and temperature-dependent drift resistance R_{drift}, as well as the bias- and temperature-dependent II mechanism. The TEOL is adopted, namely, the temperature rise over ambient $\Delta T = T - T_0$ is actually a voltage, while the dissipated power P_D is treated as a current; this allows (i) enabling the temperature sensitivity of the key physical parameters, and (ii) describing the power-temperature feedback with an electrical network. Besides the standard electrical terminals (gate, drain, source/body), the cell subcircuit is also equipped with an input node carrying the "voltage" ΔT (provided by the thermal feedback block introduced below) and with an output node offering the "current" P_D (to be fed to the thermal feedback block).

- The power-temperature feedback (i.e., the dynamic heat propagation within the structure) is described with a TEOL-based SPICE-compatible thermal feedback block, which is composed by resistances, capacitances, and controlled sources. The inputs of this block are the powers P_D dissipated by the transistor cells (represented with currents), and the outcomes are the individual (nonlinear) temperature rises ΔT (emulated with voltages).

- The thermal feedback block contains an equivalent thermal network (i.e., a purely-electrical circuit relying on the TEOL). The main contribution of the proposed approach is that this network is *automatically* constructed in the pre-processing stage by invoking the FANTASTIC tool [6] (Section 4.3). FANTASTIC receives as an input an accurate 3D FEM representation of the domain, i.e., a mesh with information about (i) the discretization into elementary cells (each corresponding to an individual heat source), (ii) material parameters, and (iii) boundary conditions, and then extracts a reduced-order model and the associated network without the need of user's expertise and COMSOL simulations; only 16 min were needed for the case study. The equivalent network accurately accounts for the self-heating of each cell and for the mutual interactions among *all* cells and describes the *linear* thermal problem. However, *nonlinear* thermal effects can be significant if the DUT is simulated under harsh conditions entailing high temperatures. In order to tackle this issue, a properly-tuned Kirchhoff's transformation [28] is used, which converts the linear temperature rises (ΔT_{lin}) into their nonlinear counterparts (ΔT) through [29]

$$\Delta T = T_0 \cdot \left[m_k + (1 - m_k) \cdot \frac{\Delta T_{lin} + T_0}{T_0} \right]^{\frac{1}{1-m_k}} - T_0 \qquad (18)$$

The calibration of the (positive) parameter m_k will be detailed in Section 4.2. Nonlinear voltage-controlled voltage sources emulating (18) are applied to the N temperature rise nodes of the equivalent thermal network; the resulting circuit is referred to as a thermal feedback block. It is worth noting that the FANTASTIC-based approach improves the strategy exploited in [7,26], where the thermal feedback block was based on Foster networks extracted in a rather long pre-processing stage from N onerous transient COMSOL simulations of the DUT (performed by activating one heat source at a time), and the thermal coupling between horizontally-far heat sources was roughly described or even neglected.

- The cell subcircuits are then connected to the thermal feedback block in a commercial circuit simulation tool (like PSPICE, LTSPICE, Eldo, ADS, SIMetrix); as a result, the whole domain under test, composed by the DUT soldered on a DBC substrate, is transformed into a *purely-electrical macrocircuit*, which suitably accounts for ET effects: the temperature, and thus the temperature-sensitive parameters, are allowed to vary during the simulation run. The solution of this macrocircuit under both static and dynamic conditions is demanded to the powerful and robust engine of the circuit simulation tool, with very low computational effort and minimized occurrence of convergence issues compared to other numerical methods.

4.1. SPICE Subcircuit and DUT Discretization into Cells

A sketch of the SPICE-compatible subcircuit for the transistor cell is represented in Figure 9, where the standard MOSFET instance at reference temperature T_0 is indicated with M_n. The additional input node feeding the "voltage" ΔT and the output node providing the "current" P_D are also represented.

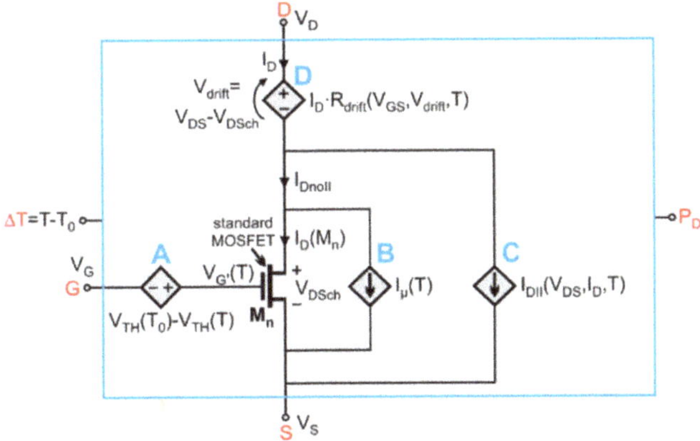

Figure 9. Sketch of the SPICE-compatible subcircuit for the transistor cell.

The negative temperature coefficient of the threshold voltage $V_{TH}(T)$ described by (3) is accounted for by using the following approach. The voltage source **A** in series with the gate adds $V_{TH}(T_0) - V_{TH}(T)$ to V_G, $V_{TH}(T)$ being given by (3), thus biasing M_n with $V_{G'} = V_G + [V_{TH}(T_0) - V_{TH}(T)]$; as a result, the *effective* overdrive voltage becomes

$$V_{G'S} - V_{TH}(T_0) = V_{GS} + [V_{TH}(T_0) - V_{TH}(T)] - V_{TH}(T_0) = V_{GS} - V_{TH}(T) \tag{19}$$

and M_n conducts the current $I_D(M_n)$ given by

$$I_D(M_n) = K(T_0) \cdot \left\{ 2[V_{GS} - V_{TH}(T)]V_{DSch} - V_{DSch}^2 \right\} \qquad V_{DSch} < V_{GS} - V_{TH}(T)$$
$$I_D(M_n) = K(T_0) \cdot [V_{GS} - V_{TH}(T)]^2 \qquad V_{DSch} \geq V_{GS} - V_{TH}(T) \tag{20}$$

The temperature dependence of the current factor $K(T)$ expressed by (4) with (5) is emulated by using the current source **B** to derive the current

$$I_\mu = I_D(M_n) \cdot \left[\left(\frac{T}{T_0} \right)^{-m(T)} - 1 \right] \tag{21}$$

Consequently, the II-unaffected current I_{Dnoll} is obtained as

$$I_{Dnoll} = I_D(M_n) + I_\mu = I_D(M_n) + I_D(M_n) \cdot \left[\left(\frac{T}{T_0} \right)^{-m(T)} - 1 \right] = I_D(M_n) \cdot \left(\frac{T}{T_0} \right)^{-m(T)} \tag{22}$$

where $m(T)$ is given by (5).

II effects are activated by adding an avalanche current $I_{DII} = \xi \cdot (I_{leak} + I_{Dnoll})$ to I_{Dnoll} through the current source **C**, $\xi = M - 1$ being bias- and temperature-dependent according to (6). Hence, the drain current I_D flowing through the drift resistance R_{drift} is given by (9).

The bias- and temperature-dependent R_{drift} described by (10) with (11) is taken into account by making use of source **D**, which imposes the voltage drop

$$V_{drift} = I_D \cdot R_{drift}(V_{GS}, V_{drift}, T).$$ (23)

Only half of the DUT was represented and simulated by exploiting its inherent symmetry. The effective active region ($\approx$5 mm^2) was partitioned into $N = 79$ cells, each with a 250×250 μm^2 area (the procedure leading to the choice of the N value will be detailed in Section 4.4). As a consequence, compared to those reported in Table 1, the values of the area-dependent parameters were scaled by dividing the current factor and the capacitances by $2 \times N$, and multiplying the resistances by $2 \times N$. The popular commercially-available OrCAD PSPICE [30] was chosen to perform circuit simulations. The main schematic (resembling the actual layout) is represented in Figure 10, along with the corresponding cell numbering and a detail of the connections between adjacent cells. The drain current of the resulting macrocircuit (an outcome of the ET simulation) has to be multiplied by 2.

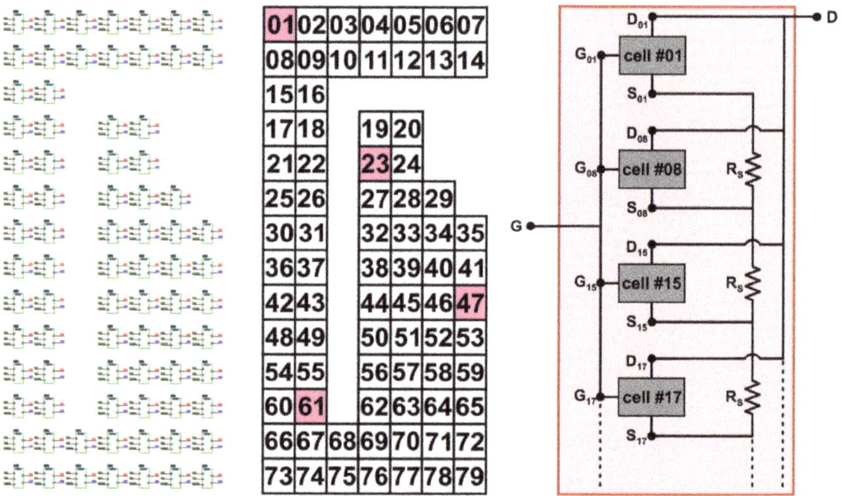

Figure 10. Discretization of the effective active area of half DUT. **Left**: PSPICE subcircuit with hierarchical blocks corresponding to the transistor cells; **center**: cell numbering, with evidenced cells of interest for the ET analysis presented in Section 5; **right**: connections between some adjacent cells.

4.2. FEM Representation of the Component under Test

Exhaustive details on the geometry and materials of the DUT (the CPMF-1200-S080B VDMOS) were taken from the datasheet and from reports available on a reverse-engineering website. The DUT was assumed to be soldered on a DBC substrate by means of a 50 μm-thick tin-platinum alloy (SnPt) layer ensuring both mechanical joint and electrical connection with the drain [31]. Figures 11 and 12 show the top view and cross-section of the assembly, which enjoys the same symmetry of the DUT. The soldering process can be automated by means of thin SnPt films, which are preformed and match the size of the die; in addition, alignment masks are typically exploited to ease their positioning. The DBC is composed by two copper (Cu) sheets (namely, bottom and top plates) with a ceramic layer placed in between them. Such a layer provides dielectric insulation to the assembly and is realized with alumina (Al$_2$O$_3$) for the proposed case study; however, alternative materials like silicon nitride (Si$_3$N$_4$) [32] and aluminum nitride (AlN) [33] are also adopted in the DBC manufacturing process. The drain contact of the DUT is soldered on the top plate, while the bottom plate ensures a good thermal interface with the 3 mm-thick Cu baseplate.

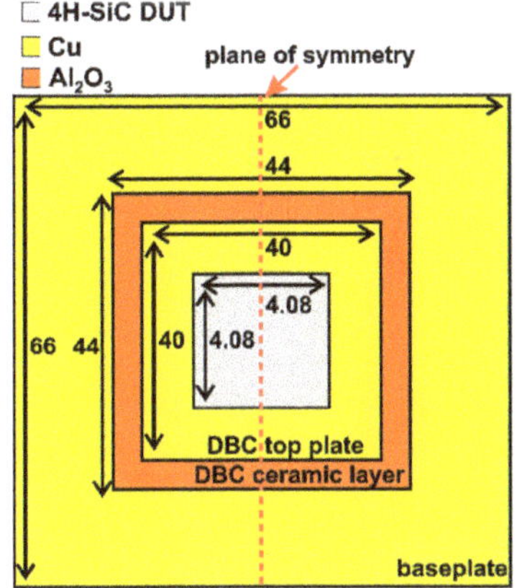

Figure 11. Sketch of the top view (not to scale) of the domain to be electrothermally simulated. All dimensions are expressed in mm.

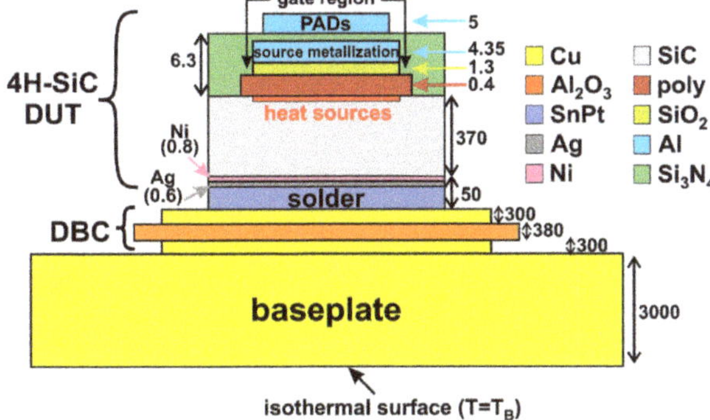

Figure 12. Sketch of the cross-section (not to scale) of the domain to be electrothermally simulated. All dimensions are expressed in µm.

The 3D domain was represented in the environment of the commercial FEM-based COMSOL Multiphysics software package [34] by means of the *in-house* routine detailed in [35], which allows *automatically* building an extremely accurate geometry (Figure 13) and performing a smart selective optimization of the tetrahedral mesh (Figure 14); this process conveniently avoids a painstakingly long and prone-to-errors manual procedure for geometry/mesh construction. The number of elements (tetrahedra) and degrees of freedom (DoFs) of the resulting grid are 3.8×10^5 and 5.2×10^5, respectively. Figure 14 plainly illustrates that the mesh is highly fine over the die, while becoming gradually coarser by moving far away from the active region.

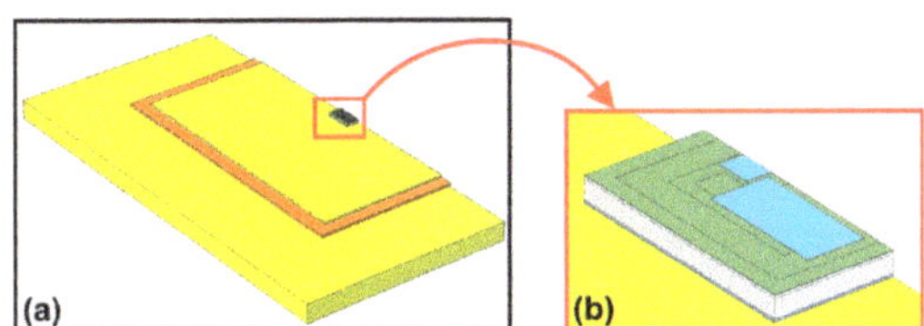

Figure 13. 3D representation of the geometry of the domain under test in the COMSOL Multiphysics environment (*draw mode*): (**a**) whole structure and (**b**) magnification of the 4H-SiC VDMOS die.

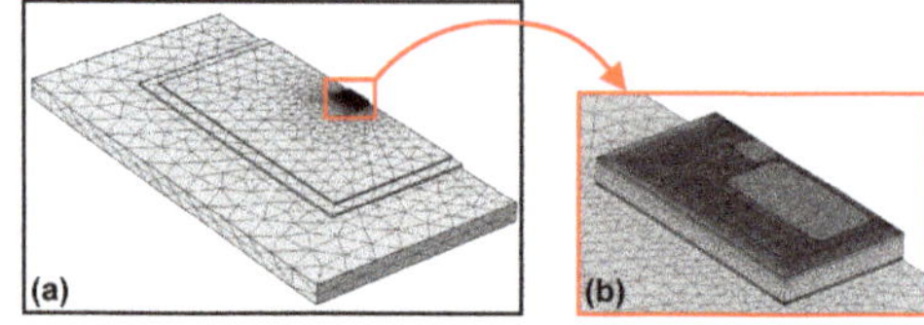

Figure 14. 3D tetrahedral mesh of the domain under test in the COMSOL Multiphysics environment (*mesh mode*): (**a**) whole structure and (**b**) magnification of the 4H-SiC VDMOS die.

As previously mentioned, all transistor cells (Figure 10) were individually associated to heat sources located over the top surface of the 4H-SiC DUT. In this representation, the heat is assumed to be dissipated over the whole effective active area, and not over the tens of thousands of individual channels; conveniently, this *unavoidable* approximation was demonstrated to be reasonable under dc and transient conditions (at least for not-too-short times) through a simulation analysis performed with TCAD Sentaurus and COMSOL. An isothermal boundary condition at $T_B = T_0$ was applied at the bottom of the assembly baseplate, whereas all other surfaces were assumed adiabatic (i.e., with zero outgoing heat flux). The material parameters adopted for the thermal simulations are listed in Table 2. Nonlinear thermal effects were accounted for by including the temperature dependences of the thermal conductivities described by

$$k(T) = k(T_0) \cdot \left(\frac{T}{T_0}\right)^{-\alpha} \tag{24}$$

$$k(T) = k(T_0) - \beta \cdot (T - T_0) \tag{25}$$

Table 2. Properties of the materials composing the assembly.

Material	$k(T_0)$ (W/mK)	c_p (J/KgK)	ρ (Kg/m^3)	α	β (W/mK^2)
4H-SiC	370 [36,37]	690 [37]	3211 [38,39] (value common to 4H-SiC and 6H-SiC)	1.29 [40]	
Al	240 [41]	905 [41]	2707 [41]		0.04 [41]
SnPt	68.8 [41]	228 [41]	7310 [41]		0.02 [41]
Ni	89.5 [41]	445 [41]	8906 [41]		0.08 [41]
Ag	427 [41]	236 [41]	10,524 [41]		0.07 [41]
poly-Si	40 [39]	920 [39]	2330 [39]		
SiO$_2$	1.38 [39]	709 [39]	2203 [39]	−0.33 [39]	
Al$_2$O$_3$	28 [39]	796 [39]	3900 [39]	1 [39]	
Cu	396.8 [41]	384 [41]	8954 [41]		0.05 [41]
Si$_3$N$_4$	18.5 [39]	787 [39]	3100 [39]	−0.33 [39]	

It must be noted that (24) applies to semiconductors and insulators, while (25) is followed by some metals.

The pre-processing calibration of parameter m_k to be used for the Kirchhoff's transformation (18) was carried out with the following procedure. The domain under test was thermally simulated with COMSOL by varying the power P_D dissipated by the whole effective active area (described with only one heat source) over a wide range (0.1 to 573.5 W for half device, the latter value leading to a peak temperature of 1400 K); the thermal conductivity dependences upon temperature were either activated (*nonlinear* conditions) or deactivated (*linear* conditions). The average temperature rise over that area was determined for both cases. Then the Kirchhoff's transformation was applied to the linear temperature rise, and m_k was adjusted so as to obtain a good agreement between the nonlinear temperature rise computed by the transformation and the realistic one evaluated by COMSOL; it was obtained that

$m_k = 0.785$. The whole process lasted less than 1 h, as each nonlinear simulation was run in about 2 mins and 24 P_D values were applied.

4.3. FANTASTIC-Based Derivation of the Equivalent Thermal Network

The FANTASTIC tool, originally introduced by some of the authors in [6], where it was applied to merely-thermal 3D simulations of a state-of-the-art gallium-arsenide (GaAs) HBT, is based on the truncated balance (TRB)-based moment matching (MM) approach to model-order reduction (MOR) developed by one of the authors in [42].

In this tool, the dynamic heat conduction problem in a semiconductor device, assumed to be *linear*, is imported from either commercial (e.g., COMSOL) or open-source codes (e.g., SALOME SMESH), comprising: the mesh discretizing the geometry, as well as the definitions of heat sources, materials, and boundary conditions. Both hexahedral and tetrahedral meshes can be used. It is possible to define arbitrary heat capacity and tensorial thermal conductivity distributions. Neumann's, Dirichlet's, or Robin's boundary conditions can be applied. Superficial (i.e., indefinitely thin) and volumetric heat sources can be taken into account.

The FEM model of the thermal problem is then assembled by FANTASTIC. In particular, the mass matrix **M** and the stiffness matrix **K** are constructed. High-order basis functions can be used: most commonly, as a trade-off between efficiency and accuracy, tetrahedral meshes and second-order basis functions are considered. The M DoFs of the temperature rise distribution, forming the M-row vector $\vartheta(t)$, are solutions of the discretized linear dynamic heat conduction problem

$$\mathbf{M}\frac{d\vartheta}{dt}(t) + \mathbf{K}\vartheta(t) = \mathbf{q}(t) \tag{26}$$

in which the power density distribution vector $\mathbf{q}(t)$ takes the form

$$\mathbf{q}(t) = \mathbf{QP}(t) \tag{27}$$

where $\mathbf{P}(t)$ is an N-row vector with the powers dissipated by N independent heat sources, and $\mathbf{Q}$ is an $M \times N$ matrix, the n-th column of which is the power density distribution vector of the n-th source, with $n = 1, \ldots, N$. The port temperature rises of the N sources form the N-row column vector $\Delta\mathbf{T}(t)$ given by [42]

$$\Delta\mathbf{T} = \mathbf{Q}^T\vartheta(t) \tag{28}$$

As typical for MOR approaches, an $M \times \hat{M}$ matrix **V** with $\hat{M} \ll M$ is defined, which allows approximating $\vartheta(t)$ by means of a reduced number $\hat{M}$ of DoFs forming the $\hat{M}$-vector $\hat{\vartheta}(t)$, so that

$$\vartheta(t) = \mathbf{V}\hat{\vartheta}(t) \tag{29}$$

The **V** matrix is used to project the heat conduction discretized problem (26)–(28) with the Galerkin's method, thus deriving a DCTM in the form

$$\hat{\mathbf{M}}\frac{d\hat{\vartheta}}{dt}(t) + \hat{\mathbf{K}}\hat{\vartheta}(t) = \hat{\mathbf{q}}(t) \tag{30}$$

being

$$\hat{\mathbf{q}}(t) = \hat{\mathbf{G}}\mathbf{P}(t) \tag{31}$$

$$\Delta\mathbf{T}(t) = \hat{\mathbf{G}}^T\hat{\vartheta}(t) \tag{32}$$

in which

$$\hat{\mathbf{M}} = \mathbf{V}^T\mathbf{M}\mathbf{V} \tag{33}$$

$$\hat{\mathbf{K}} = \mathbf{V}^T\mathbf{K}\mathbf{V} \tag{34}$$

are $\hat{M}$-order matrices, and

$$\hat{G} = V^T Q \tag{35}$$

is an $\hat{M} \times N$ matrix. The **V** matrix is determined by the algorithm reported below.

Algorithm 1: DCTM extraction

Set **V**:=0
for each independent heat source n=1, ..., N **do**

1   Compute complex frequency values σ_p with p=1, ..., P_n

   for p=1, ..., P_n **do**

     Set $\widetilde{\Theta}_p$ =0

     if p>1 **then**

2   Solve compact model $\mathcal{C}_{p-1}$ given by (38) for $\widehat{\Theta}_p$

       Compute $\widetilde{\Theta}_p = V_{p-1}\widehat{\Theta}_p$ approximating Θ_p

3   Solve (37) for Θ_p using $\widetilde{\Theta}_p$ as initial estimation in the iterative technique
4   Generate matrix V_p spanning the columns of V_{p-1} and Θ_p
5   Generate a compact model $\mathcal{C}_p$, projecting (26)–(28) onto V_p

   Append the columns of V_{P_n} to matrix **V**

At line 1 of the algorithm, the values σ_p, with $p = 1, \ldots, P_n$, are automatically determined as a function of the desired relative error parameter ε as follows. First, the real positive quantities $\lambda_n < \Lambda_n$ are properly estimated for the heat conduction problem with respect to the power impulse thermal response of the current p-th heat source. Next, the value P_n is determined as the smallest integer such that

$$4\exp\left(-P_n\pi^2/\log(4/k')\right) \leq \varepsilon \tag{36}$$

being $k' = \lambda_n/\Lambda_n$. It is then set

$$\sigma_p = \Lambda_n \mathrm{dn}\left(\frac{2p-1}{2P_n}K, k\right)$$

with $p = 1, \ldots, P_n$, in which K is the complete elliptic integral of the first kind of modulus k, and dn is the homonymous elliptic function of modulus $k = \sqrt{1 - k'^2}$.

At line 3, the temperature response to the n-th heat source is solved in the complex frequency domain at the frequency value σ_p. Thus, equation

$$(\sigma_p M + K)\Theta_p = Qe_n \tag{37}$$

is solved for Θ_p, e_n being an N-row vector having all zero elements but one at the n-th row. Since the complex frequency values are real positive, the coefficient matrices of the resulting linear systems are symmetric positive definite, and the most efficient multigrid iterative solvers can be used for their solution.

At line 4, the V_p matrix is achieved by appending to the columns of V_{p-1} a vector derived orthogonalizing Θ_p with respect to the columns of V_{p-1}.

At line 5, the DCTM C_p is determined proceeding as in (30)–(32), matrix **V** in (33)–(35) being substituted by matrix V_p.

At line 2, the temperature response Θ_p in the complex frequency domain due to the n-th independent heat source is estimated at σ_p using the DCTM C_{p-1}. To this aim, equation

$$(\sigma_p \hat{M}_{p-1} + \hat{K}_{p-1})\hat{\Theta}_p = \hat{Q}_{p-1}e_n \tag{38}$$

is solved for $\hat{\Theta}_p$. This vector is used to determine the approximation $\widetilde{\Theta}_p = V_{p-1}\hat{\Theta}_p$. At line 3 of the algorithm, such estimation of Θ_p is used to speed up the solution of (37) by setting the initial estimation of the solution.

It is noted that for the n-th heat source, with $n = 1, \ldots, N$, the complex frequency values σ_p, with $p = 1, \ldots, P_n$, are sorted in decreasing order. In this way, the linear systems introduced at line 3 of the algorithm are solved from the least to the most onerous ones in terms of CPU time. The computational burden is drastically reduced by this proper ordering, since the iterative solver can start from an estimate of the solution found in the previous step, which becomes increasingly accurate.

The DCTM achieved by this algorithm has dimension $\hat{M} = P_1 + P_2 + \ldots + P_N$. Let $\mathbf{Z}(t)$ and $\hat{\mathbf{Z}}(t)$ be the N-th order power impulse thermal response [K/W] matrices of the discretized heat conduction problem (26)–(28) and of the DCTM, respectively. As shown in [43], it results in

$$\|\mathbf{Z}(t) - \hat{\mathbf{Z}}(t)\|_{\mathcal{H}_2} \leq 2\varepsilon \|\mathbf{Z}(t)\|_{\mathcal{H}_2}$$

being

$$\| \mathbf{Z}(t) \|_{\mathcal{H}_2} = \sqrt{\int_0^{+\infty} \mathrm{tr}\left(\mathbf{Z}^T(t)\mathbf{Z}(t)\right)dt}$$

the $\mathbf{Z}(t)$ Hankel norm. Similar results can be extended from the time to the frequency domain and can be stated for the whole spatial-temporal temperature distributions due to power impulses [43]. All these results provide a strong theoretical guarantee for the convergence of the method. As a consequence of (36), this convergence is very fast, being exponential with respect to the numbers P of σ_p, with $p = 1, \ldots, P_n$. As a result, in practice accurate DCTMs can always be obtained with small space-state dimensions, $\hat{M}$.

It is now observed that by solving at limited cost the generalized eigenvalue problem

$$\hat{\mathbf{U}}^T\hat{\mathbf{M}}\hat{\mathbf{U}} = \mathbf{I}_{\hat{M}}$$

$$\hat{\mathbf{U}}^T\hat{\mathbf{K}}\hat{\mathbf{U}} = \hat{\mathbf{\Lambda}}$$

having as unknown the $\hat{M}$-order matrix $\hat{\mathbf{U}}$, in which $\hat{\mathbf{\Lambda}}$ is an $\hat{M}$-order diagonal matrix, and introducing the change of variables

$$\hat{\vartheta}(t) = \hat{\mathbf{U}}\hat{\xi}(t)$$

the DCTM equations (30)–(32) are transformed into

$$\frac{d\hat{\xi}}{dt}(t) + \hat{\mathbf{\Lambda}}\hat{\xi}(t) = \hat{\mathbf{\Gamma}}\mathbf{P}(t) \tag{39}$$

$$\Delta\mathbf{T}(t) = \hat{\mathbf{\Gamma}}^T\hat{\xi}(t) \tag{40}$$

where $\hat{\mathbf{\Gamma}}$ is the $\hat{M} \times N$ matrix $\hat{\mathbf{V}}^T\mathbf{G}$. The temperature rise distribution is then reconstructed as

$$\vartheta(t) = \Xi\hat{\xi}(t) \tag{41}$$

$\Xi = \mathbf{V}\hat{\mathbf{U}}$ being an $M \times \hat{M}$ matrix like $\mathbf{V}$.

Equations (39) and (40) can be interpreted as the equations ruling the equivalent network sketched in Figure 15.

Such a network, which can in principle describe the thermal behavior of *any* electronic component, is particularly suited to be solved by means of nodal analysis in SPICE-like simulators, since all elements are voltage-controlled current sources, and thus the number of variables is limited to $\hat{M}$. The topology is general and can be implemented into *any* circuit simulator.

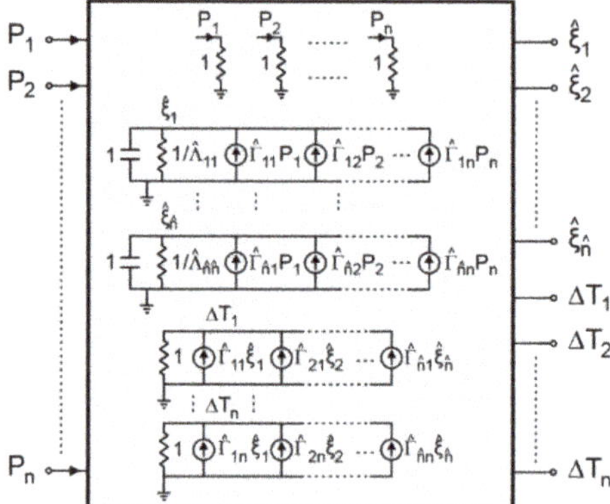

Figure 15. DCTM equivalent thermal network (after Codecasa et al. [6]).

4.4. Pre-Processing Evaluation of N and ε

During the pre-processing stage, two key parameters have to be chosen, namely, the number of elementary cells N (=79, as mentioned in Section 4.1) and the relative error parameter ε, the latter needed for the generation of the equivalent thermal network with FANTASTIC and set to 10^{-3}. For the particular component under test, this discretization and this error were found to be a good trade-off between accuracy and CPU time needed for the ET simulation, as evaluated in an additional pre-processing analysis where some N and ε values were tested; considering much higher N and/or much lower ε leads to a marginal accuracy improvement paid with a significant increase in CPU time.

4.5. Construction of the Macrocircuit

The macrocircuit representing the ET behavior of the packaged DUT was constructed in the PSPICE environment as follows. The linear equivalent thermal network, provided in the form of a netlist, was enriched with N nonlinear voltage-controlled voltage sources to account for the Kirchhoff's transformation (18), and the thermal feedback block was thus obtained. It must be remarked that nonlinear thermal effects could in principle be *accurately* described by extracting a *fully nonlinear* equivalent thermal network with a variant of the FANTASTIC tool [11]. However, the complexity of the nonlinear network grows with the discretization N much more rapidly than that of the linear counterpart (used in this paper); for $N = 79$, such a network would be composed by about 32×10^6 elements, with insurmountable memory-storage problems. As a consequence, the adoption of a calibrated Kirchhoff's transformation represents the only viable strategy to get accurate enough results. The ΔT and P_D nodes of the subcircuits (the individual transistor cells) were connected to the thermal feedback block. As shown in Figure 10c, all the gate terminals of the subcircuits were shorted together, as well as the drain and source ones, and it is possible to activate an electrical network to include the de-biasing over the source pad. A simplified scheme of the adopted strategy is shown in Figure 16.

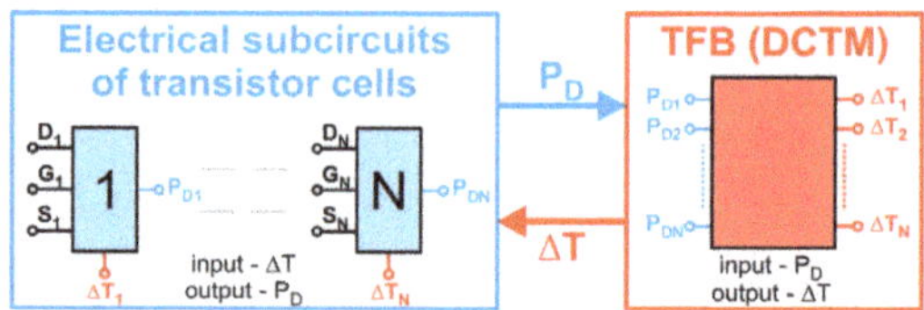

Figure 16. Schematic representation of the proposed strategy to perform a fully coupled ET analysis in a circuit simulation tool: the feedback loop between the electrical circuit (**left**) and the thermal feedback block (TFB, **right**) relying on the DCTM-based equivalent network is highlighted.

After the circuit simulation run, the whole spatial-temporal temperature rise distribution can be reconstructed in a post-processing stage at negligible computational cost and memory storage using (41).

5. Static and Dynamic Electrothermal Simulations

The whole pre-processing activity, involving isothermal measurements, optimization of the model parameters, geometry/mesh construction in COMSOL, proper domain discretization, equivalent network extraction with FANTASTIC, calibration of the Kirchhoff's transformation, and macrocircuit generation, can last a few days, after which *any* analysis can be performed in short times on the device of interest.

In particular, the macrocircuit was adopted to perform many PSPICE simulations of the DUT with DBC substrate in both static (dc) and dynamic conditions on a PC with an Intel Core i7-7700 (3.60 GHz) CPU and equipped with a 16 GB RAM. Unfortunately, experimental data for accuracy validation were not available for the examined component. On the other hand, the approach benefits from (i) a careful calibration of the transistor model from experimental data; (ii) a very accurate domain representation in COMSOL aided by the *in-house* routine mentioned in Section 4.2; (iii) an automated (and unaffected by user's errors) extraction of the linear equivalent network through FANTASTIC; (iv) a smart pre-processing calibration of the Kirchhoff's transformation. Hence, the only source of error can be induced by the degree of uncertainty concerning the thermal conductivities of the involved materials, which however affects any ET simulation approach.

First, the I_D–V_{DS} output characteristics were determined with a V_{DS} step amounting to 0.1 V under isothermal (at T_0) and ET conditions. Isothermal conditions were obtained by deactivating the thermal feedback block. The CPU time needed to simulate a single ET characteristic was nearly 100 s. Results are reported in Figure 17, which also shows the temperature rise above T_0 averaged over the effective active area (ΔT_{av}). It can be inferred that the simulation runs were stopped as ΔT_{av} reached 500 K.

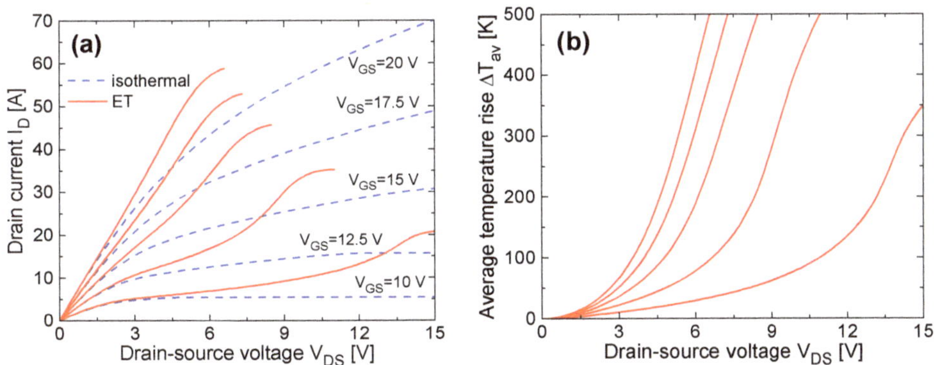

Figure 17. (**a**) I_D–V_{DS} output characteristics determined with the proposed approach under isothermal (dashed blue lines) and ET (solid red) conditions; (**b**) temperature rise $\Delta T_{av} = T_{av} - T_0$ corresponding to the ET conditions, T_{av} being the temperature averaged over the effective active area.

Afterward, SC tests were simulated. Such tests involve large power dissipation, and are typically used to quantify the device robustness under harsh and abnormal events (see e.g., [7,23,25,44–47], all focused on SiC MOSFETs). In the SC experiment, the DUT is first biased in the OFF state with a given supply voltage applied to the drain, and then turned on with a single gate pulse (a gate resistance R_{GATE} of 50 Ω was considered). The knowledge of the whole temperature distribution over the active area is important, since the value and position of the temperature peak are needed for reliability considerations. The effects of many combinations of gate and supply voltages were examined. Simulation runs were very fast: a single test required about 300 s with a fine time discretization.

The total drain current I_D conducted by the DUT vs. time is shown in Figure 18a for all the analyzed cases, while Figure 18b illustrates the corresponding temperature rises ΔT_{av}. The first figure reveals that I_D first grows due to the strong positive temperature coefficient induced by (i) the reduction in threshold voltage and (ii) the mobility increase (for the lowered Coulomb scattering), and then drops since the negative temperature coefficient triggered by the acoustic-phonon scattering dominates (thermally stable behavior) as all the traps have released electrons [7,25].

The temperature rises ΔT over cells #01, #23, #47, #61 (identified in the layout of Figure 10) are reported in Figure 18c for two cases, namely, $V_{GS} = 10$ V/$V_{DD} = 200$ V and $V_{GS} = 20$ V/$V_{DD} = 200$ V. From the inspection of the waveforms, it is found that a pronounced temperature nonuniformity takes place in the first case (the inner cell #47 suffers from a ΔT two times higher than that of the top-corner cell #01), which can be explained as follows. The milder bias conditions ($V_{GS} = 10$ V) allowed the device to safely undergo the test for a longer period, within which the heat had enough time to significantly spread, thereby favoring a stronger impact of the mutual thermal interactions and the consequent exacerbation of temperature gradients.

The whole temperature field in the domain was determined at chosen time instants from the DCTM generated with FANTASTIC in a post-processing step. More specifically, points A ($t = 25.5$ μs, $V_{GS} = 20$ V/$V_{DD} = 200$ V), B ($t = 338$ μs, $V_{GS} = 20$ V/$V_{DD} = 50$ V), and C ($t = 14$ ms, $V_{GS} = 10$ V/$V_{DD} = 50$ V) identified in Figure 18a,b were selected, which approximately share the same ΔT_{av} value, i.e., 500 K. The computed temperature rise maps are shown in Figure 18d (top view) and Figure 18e (side view). As can be seen, despite the same ΔT_{av},

- point A, which falls at a time instant close to the beginning of the test, is endowed with a uniform and superficial temperature field, since the heat is still confined in the top active region close to the generation area;
- the temperature distributions in B and C are increasingly uneven, which is again ascribable to the much longer stress times. In particular, C suffers from a severe temperature focusing over the innermost DUT area, and—as witnessed by the side view—the downward heat had enough time to reach and hit the DBC.

Lastly, the macrocircuit was used to simulate two UIS tests, which are commonly adopted to evaluate the maximum amount of avalanche energy sustainable by the device [8,44]. In the UIS experiment, a load inductor L tied to the drain is first ramped up to a desired current by keeping the DUT in linear mode for a time t_{ON}; then the DUT is turned off and brought into avalanche by the inductor, which preserves the current continuity. The tests will be hereinafter denoted as case #1 and #2, the specifics of which are as follows:

- case #1: $V_{DD} = 300$ V, $L = 4.6$ mH, $R_{GATE} = 15$ Ω
- case #2: $V_{DD} = 600$ V, $L = 12$ mH, $R_{GATE} = 15$ Ω

In both cases, a gate voltage equal to 20 V was applied for $t_{ON} = 200$ μs, and then lowered to 0. Again, the time elapsed by PSPICE for a single test was about 300–400 s. Concerning case #1, Figure 19a shows the drain current I_D and the drain-source voltage V_{DS} vs. time in the span 190 to 280 μs, while Figure 19b illustrates the temperature rises of cells #01, #23, #47, #61. An almost uniform temperature distribution was found, as witnessed by the map taken at time instant $t^* = 220$ μs, where the dynamic temperature averaged over the active area peaks (Figure 19c).

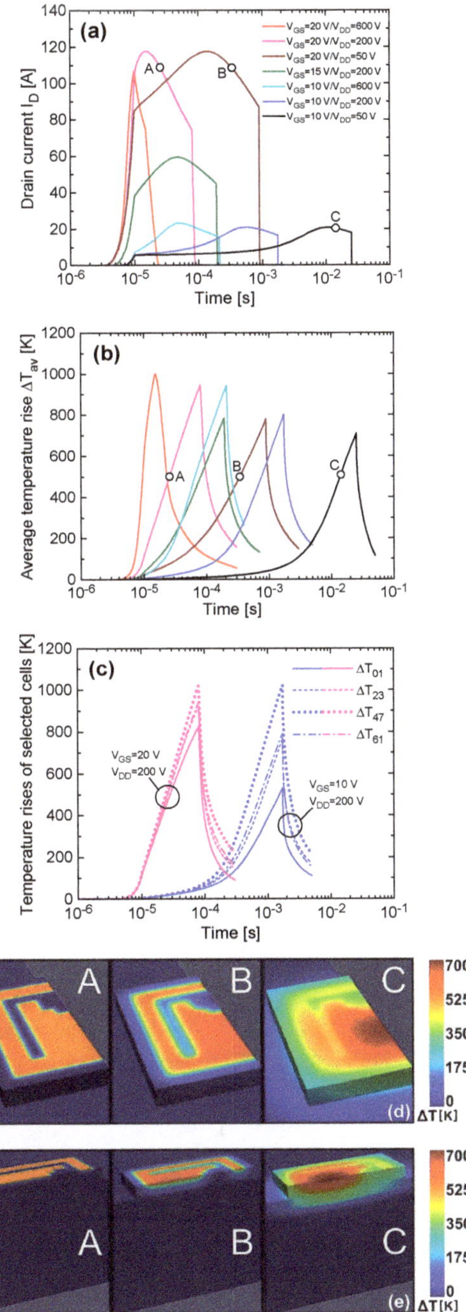

Figure 18. Simulated SC test: (**a**) drain current I_D vs. time for seven V_{GS}/V_{DD} combinations; (**b**) corresponding temperature rises averaged over the effective active area; (**c**) temperature rises of the individual cells highlighted in Figure 10 for 2 cases; (**d**) top and (**e**) side views of the 3D temperature rise maps determined at points A, B, and C shown in (**a**) and (**b**).

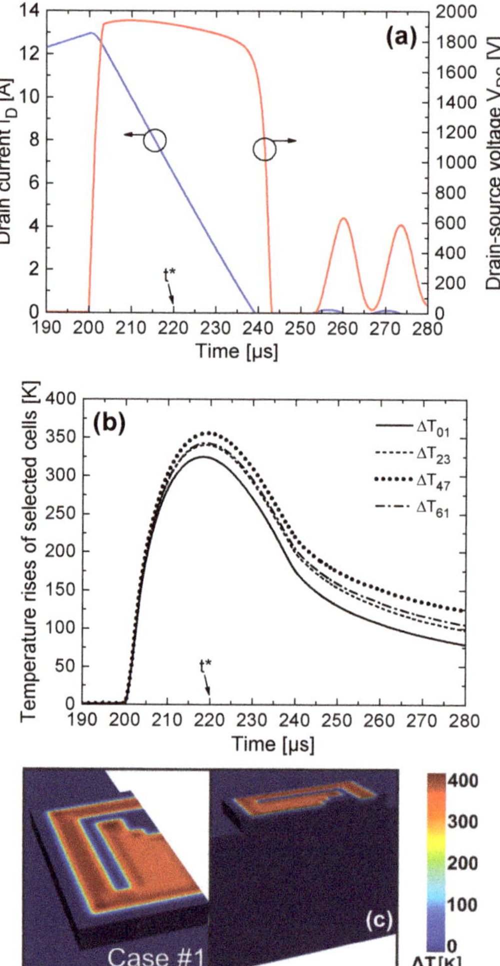

Figure 19. Simulated UIS test, case #1: (**a**) drain current I_D and drain-source voltage V_{DS} against time; (**b**) temperature rises of the individual cells identified in Figure 10; (**c**) top and side views of the temperature rise maps calculated at time instant t* = 220 µs shown in (**a**) and (**b**).

In case #2, despite the lower drain current $I_D \approx V_{DD} \cdot t_{ON}/L$ before turn-off (10 A instead of 13 A obtained for case #1), the average temperature reaches higher values due the longer discharging transient, in turn dictated by the higher L, since

$$\frac{dI_D}{dt} = -\frac{BV_{DS}(T) - V_{DD}}{L} \tag{42}$$

Results are shown in Figure 20. As can be inferred from Figure 20b, which reports the temperature rises of the selected cells, and from Figure 20c, which depicts the post-processing temperature maps at t* = 240 µs, a slightly more nonuniform temperature field occurs with respect to case #1.

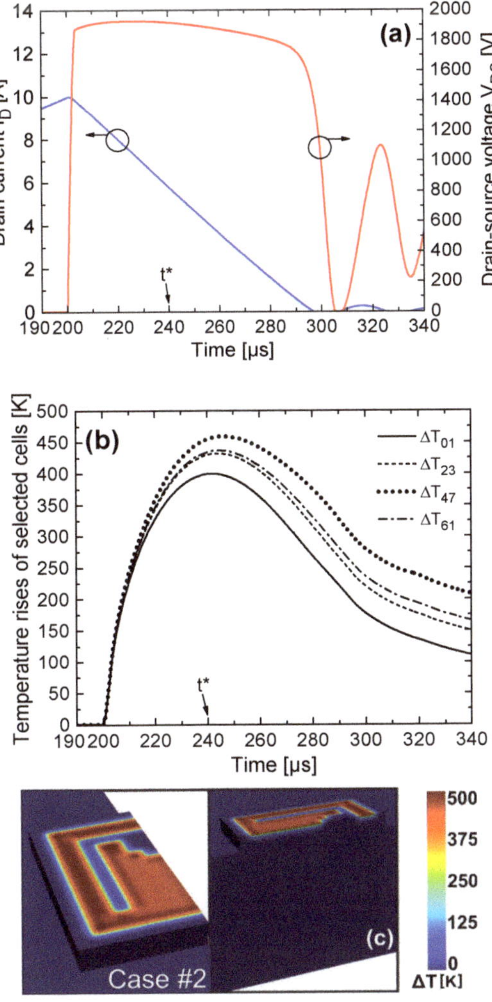

Figure 20. Simulated UIS test, case #2: (**a**) drain current I_D and drain-source voltage V_{DS} vs. time; (**b**) temperature rises of the individual cells highlighted in Figure 10; (**c**) top and side views of the temperature rise maps calculated at time instant t* = 240 µs identified in (**a**) and (**b**).

6. Conclusions

In this paper, an advanced circuit-based approach for the static and dynamic electrothermal simulation of multicellular SiC power MOSFETs has been proposed, which—differently from the strategies encountered in the literature—seems to represent a good trade-off between accuracy and efficiency. The device is discretized into a chosen number of elementary cells (heat sources) and turned into a purely-electrical macrocircuit, where (i) the cells are described with subcircuits accounting for the key and harmful influence of SiC/SiO$_2$ interface traps, and (ii) the power-temperature feedback is modeled with an equivalent thermal network. This network is obtained through a *fully automated* process: first, a 3D mesh representing the device is generated in COMSOL with the support of an *in-house* routine that elaborates the layout files and makes use of further information about thickness of layers, position/shape of the heat sources, material parameters, and boundary conditions; then, such a mesh is provided as an input to the FANTASTIC tool, which derives a dynamic compact thermal model of the device and the related equivalent network. The macrocircuit can be solved in the environment of *any* SPICE-like simulator with short CPU time and unlikely occurrence of convergence issues. The effectiveness and efficiency of the proposed strategy have been verified on a 1200 V, 50 A multicellular 4H-SiC VDMOS soldered on a DBC package. It has been shown that the simulation of short-circuit and unclamped inductive switching tests requires only 300–400 s on a normal PC, despite the critical conditions and the fine time discretization, and that potentially-dangerous temperature gradients/hogging can be easily identified. It can be concluded that the proposed approach can be helpful for industry engineers who are in charge of optimizing the thermal design of multicellular power devices in *any* technology.

Author Contributions: Methodology, V.d.; Software, V.d., L.C., A.P.C., C.S.; Validation, V.d.; Writing–Original Draft Preparation, V.d.; Writing–Review & Editing, V.d.; Supervision, V.d. All authors have read and agreed to the published version of the manuscript.

Funding: This research received no external funding.

Acknowledgments: The funding for the Ph.D. activity of Ciro Scognamillo was generously donated by the Rinaldi family in the memory of Niccolò Rinaldi, a bright Professor and Researcher of University of Naples Federico II, prematurely passed away in 2018.

Conflicts of Interest: The authors declare no conflict of interest.

Nomenclature

AlN	aluminum nitride
Al_2O_3	alumina
Cu	copper
GaAs	gallium arsenide
Si	silicon
SiC	silicon carbide
SiO_2	silicon dioxide (or oxide)
Si_3N_4	silicon nitride
SnPt	tin-platinum alloy
DBC	direct-bonded copper
DCTM	dynamic compact thermal model
DoF	degree of freedom
DUT	device under test
ET	electrothermal
FANTASTIC	FAst Novel Thermal Analysis Simulation Tool for Integrated Circuits
FEM	finite-element method
HBT	heterojunction bipolar transistor
II	impact ionization
ILS	inductive load switching
MM	moment matching
MOR	model-order reduction
MOSFET	metal oxide semiconductor field effect transistor
SC	short circuit
TEOL	thermal equivalent of the Ohm's law
TRB	truncated balance
UIS	unclamped inductive switching
VDMOS	vertical double-diffused MOS

References

1. Östling, M.; Ghandi, R.; Zetterling, C.-M. SiC power devices—Present status, applications and future perspectives. In Proceedings of the IEEE International Symposium on Power Semiconductor Devices and IC's (ISPSD), San Diego, CA, USA, 23–26 May 2011; pp. 10–15.
2. De Falco, G.; Riccio, M.; Romano, G.; Maresca, L.; Irace, A.; Breglio, G. ELDO-COMSOL based 3D electro-thermal simulations of power semiconductor devices. In Proceedings of the IEEE Annual SEMIconductor THERMal measurement, modeling, and management symposium (SEMI-THERM), San Jose, CA, USA, 9–13 March 2014; pp. 35–40.
3. Chvála, A.; Donoval, D.; Marek, J.; Príbytný, P.; Molnár, M.; Mikolášek, M. Fast 3-D electrothermal device/circuit simulation of power superjunction MOSFET based on SDevice and HSPICE interaction. *IEEE Trans. Electron Devices* **2014**, *61*, 1116–1122. [CrossRef]
4. Košel, V.; de Filippis, S.; Chen, L.; Decker, S.; Irace, A. FEM simulation approach to investigate electro-thermal behavior of power transistors in 3-D. *Microelectron. Reliab.* **2013**, *53*, 356–362. [CrossRef]
5. Pfost, M.; Boianceanu, C.; Lohmeyer, H.; Stecher, M. Electrothermal simulation of self-heating in DMOS transistors up to thermal runaway. *IEEE Trans. Electron Devices* **2013**, *60*, 699–707. [CrossRef]
6. Codecasa, L.; d'Alessandro, V.; Magnani, A.; Rinaldi, N.; Zampardi, P.J. FAst Novel Thermal Analysis Simulation Tool for Integrated Circuits (FANTASTIC). In Proceedings of the IEEE international workshop on THERMal INvestigations of ICs and systems (THERMINIC), Greenwich, London, UK, 24–26 September 2014.
7. d'Alessandro, V.; Magnani, A.; Riccio, M.; Breglio, G.; Irace, A.; Rinaldi, N.; Castellazzi, A. SPICE modeling and dynamic electrothermal simulation of SiC power MOSFETs. In Proceedings of the IEEE International Symposium of Power Semiconductor Devices and IC's (ISPSD), Waikoloa, HI, USA, 15–19 June 2014.
8. Fayyaz, A.; Romano, G.; Urresti, J.; Riccio, M.; Castellazzi, A.; Irace, A.; Wright, N. A comprehensive study of the avalanche breakdown robustness of silicon carbide power MOSFETs. *Energies* **2017**, *10*, 452. [CrossRef]
9. Riccio, M.; d'Alessandro, V.; Romano, G.; Maresca, L.; Breglio, G.; Irace, A. A temperature-dependent SPICE model of SiC power MOSFETs for within and out-of-SOA simulations. *IEEE Trans. Power Electron.* **2018**, *33*, 8020–8029. [CrossRef]
10. Mantooth, H.A.; Peng, K.; Santi, E.; Hudgins, J.L. Modeling of wide bandgap semiconductor devices–Part I. *IEEE Trans. Electron Devices* **2015**, *62*, 423–433. [CrossRef]
11. Codecasa, L.; d'Alessandro, V.; Magnani, A.; Irace, A. Circuit-based electrothermal simulation of power devices by an ultrafast nonlinear MOS approach. *IEEE Trans. Power Electron.* **2016**, *31*, 5906–5916. [CrossRef]
12. Rinaldi, N.; d'Alessandro, V. Theory of electrothermal behavior of bipolar transistors: Part III–Impact ionization. *IEEE Trans. Electron Devices* **2006**, *53*, 1683–1697. [CrossRef]
13. Ren, Y.; Xu, M.; Zhou, J.; Lee, F.C. Analytical loss model of power MOSFET. *IEEE Trans. Power Electron.* **2006**, *21*, 310–319.
14. Baliga, B.J. *Modern Power Devices*; Wiley: New York, NY, USA, 1987.
15. Arora, N. *MOSFET Modeling for VLSI Simulation: Theory and Practice*; World Scientific Publishing Co. Pte. Ltd.: Singapore, 2007.
16. Ólafsson, H.Ö.; Gudjónsson, G.; Allerstam, F.; Sveinbjörnsson, E.Ö.; Rödle, T.; Jos, R. Stable operation of high mobility 4H-SiC MOSFETs at elevated temperatures. *Electron. Lett.* **2005**, *41*, 825–826.
17. Wang, J.; Zhao, T.; Li, J.; Huang, A.Q.; Callanan, R.; Husna, F.; Agarwal, A. Characterization, modeling, and application of 10-kV SiC MOSFETs. *IEEE Trans. Electron Devices* **2008**, *55*, 1798–1806. [CrossRef]
18. Chen, S.; Cai, C.; Wang, T.; Guo, Q.; Sheng, K. Cryogenic and high temperature performance of 4H-SiC power MOSFETs. In Proceedings of the Annual IEEE Applied Power Electronics Conference and Exposition (APEC), Long Beach, CA, USA, 17–21 March 2013; pp. 207–210.
19. Spirito, P.; Breglio, G.; d'Alessandro, V.; Rinaldi, N. Thermal instabilities in high current power MOS devices: Experimental evidence, electro-thermal simulations and analytical modeling. In Proceedings of the IEEE International Conference on MIcroELectronics (MIEL), Niš, Yugoslavia, 12–15 May 2002; pp. 23–30.
20. Spirito, P.; Breglio, G.; d'Alessandro, V.; Rinaldi, N. Analytical model for thermal instability of low voltage power MOS and S.O.A. in pulse operation. In Proceedings of the IEEE International Symposium on Power Semiconductor Devices and IC's (ISPSD), Santa Fe, NM, USA, 3–7 June 2002; pp. 269–272.

21. Pérez-Tomás, A.; Brosselard, P.; Godignon, P.; Millán, J.; Mestres, N.; Jennings, M.R.; Covington, J.A.; Mawby, P.A. Field-effect mobility temperature modeling of 4H-SiC metal-oxide-semiconductor transistors. *J. Appl. Phys.* **2006**, *100*, 114508. [CrossRef]

22. Cheng, L.; Agarwal, A.K.; Dhar, S.; Ryu, S.-H.; Palmour, J.W. Static performance of 20 A, 1200 V 4H-SiC power MOSFETs at temperatures of −187 °C to 200 °C. *J. Electron. Mater.* **2012**, *41*, 910–914. [CrossRef]

23. Huang, X.; Wang, G.; Li, Y.; Huang, A.Q.; Baliga, B.J. Short-circuit capability of 1200V SiC MOSFET and JFET for fault protection. In Proceedings of the Annual IEEE Applied Power Electronics Conference and Exposition (APEC), Long Beach, CA, USA, 17–21 March 2013; pp. 197–200.

24. TCAD Sentaurus User's Manual, Synopsis, 2015. Available online: https://www.synopsys.com/silicon/tcad/device-simulation/sentaurus-device.html (accessed on 25 May 2020).

25. Castellazzi, A.; Funaki, T.; Kimoto, T.; Hikihara, T. Thermal instability effects in SiC power MOSFETs. *Microelectron. Reliab.* **2012**, *52*, 2414–2419. [CrossRef]

26. d'Alessandro, V.; Magnani, A.; Riccio, M.; Iwahashi, Y.; Breglio, G.; Rinaldi, N.; Irace, A. Analysis of the UIS behavior of power devices by means of SPICE-based electrothermal simulations. *Microelectron. Reliab.* **2013**, *53*, 1713–1718. [CrossRef]

27. Bazzano, G.; Cavallaro, D.G.; Greco, G.; Grimaldi, A.; Rinaudo, S. Stress and reliability of power devices: An innovative thermal analysis approach to predict a device's lifetime. In Proceedings of the International Exhibition & Conference for Power Electronics, Intelligent Motion, Power Quality (PCIM Europe), Nuremberg, Germany, 17–19 May 2011; pp. 117–123.

28. Joyce, W.B. Thermal resistance of heat sinks with temperature-dependent conductivity. *Solid State Electron.* **1975**, *18*, 321–322. [CrossRef]

29. Poulton, K.; Knudsen, K.L.; Corcoran, J.J.; Wang, K.-C.; Pierson, R.L.; Nubling, R.B.; Chang, M.-C.F. Thermal design and simulation of bipolar integrated circuits. *IEEE J. Solid State Circuits* **1992**, *27*, 1379–1387. [CrossRef]

30. PSPICE User's Manual, Cadence OrCAD 16.5, 2011. Available online: https://www.orcad.com/ (accessed on 25 May 2020).

31. Li, H.; Munk-Nielsen, S.; Bęczkowski, S.; Wang, X. A novel DBC layout for current imbalance mitigation in SiC MOSFET multichip power modules. *IEEE Trans. Power Electron.* **2016**, *31*, 8042–8045. [CrossRef]

32. Suganuma, K.; Kim, S. Ultra heat-shock resistant die attachment for silicon carbide with pure zinc. *IEEE Electron Device Lett.* **2010**, *31*, 1467–1469. [CrossRef]

33. Catalano, A.P.; Scognamillo, C.; Castellazzi, A.; d'Alessandro, V. Optimum thermal design of high-voltage double-sided cooled multi-chip SiC power modules. In Proceedings of the IEEE International Workshop on THERMal INvestigations of ICs and Systems (THERMINIC), Lecco, Italy, 25–27 September 2019.

34. COMSOL Multiphysics User's Guide, Release 5.2A, 2016. Available online: https://www.comsol.it/ (accessed on 25 May 2020).

35. d'Alessandro, V.; Catalano, A.P.; Codecasa, L.; Zampardi, P.J.; Moser, B. Accurate and efficient analysis of the upward heat flow in InGaP/GaAs HBTs through an automated FEM-based tool and Design of Experiments. *Int. J. Numer. Model. Electron. Netw. Devices Fields* **2019**, *32*.

36. Harris, G.L. *Properties of Silicon Carbide*; INSPEC, the Institution of Electrical Engineers: London, UK, 1995.

37. Goldberg, Y.; Levinshtein, M.E.; Rumyantsev, S.L. Silicon Carbide. In *Properties of Advanced Semiconductor Materials: GaN, AlN, InN, BN, SiC, SiGe*; Levinshtein, M.E., Rumyantsev, S.L., Shur, M.S., Eds.; John Wiley & Sons, Inc.: New York, NY, USA, 2001; Volume 5, pp. 93–148.

38. Gomes de Mesquita, A.H. Refinement of the crystal structure of SiC type 6H. *Acta Crystallogr.* **1967**, *23*, 610–617. [CrossRef]

39. Palankovski, W.; Quay, R. *Analysis and Simulation of Heterostructure Devices*; Springer: New York, NY, USA, 2004.

40. Joshi, R.P.; Neudeck, P.G.; Fazi, C. Analysis of the temperature dependent thermal conductivity of silicon carbide for high temperature applications. *J. Appl. Phys.* **2000**, *88*, 265–269. [CrossRef]

41. Lienhard, I.V.; Lienhard, H. *A Heat Transfer Textbook*; Phlogiston Press: Cambridge, MA, USA, 2008.

42. Codecasa, L.; D'Amore, D.; Maffezzoni, P. Compact modeling of electrical devices for electrothermal analysis. *IEEE Trans. Circuits Syst. I Fundam. Theory Appl.* **2003**, *50*, 465–476. [CrossRef]

43. Codecasa, L.; Catalano, A.P.; d'Alessandro, V. A *Priori* error bound for moment matching approximants of thermal models. *IEEE Trans. Compon. Packag. Manuf. Technol.* **2019**, *9*, 2383–2392. [CrossRef]

44. Fayyaz, A.; Yang, L.; Castellazzi, A. Transient robustness testing of silicon carbide (SiC) power MOSFETs. In Proceedings of the European Conference on Power Electronics and Applications (EPE), Lille, France, 2–6 September 2013.

45. Nguyen, T.-T.; Ahmed, A.; Thang, T.V.; Park, J.-H. Gate oxide reliability issues of SiC MOSFETs under short-circuit operation. *IEEE Trans. Power Electron.* **2015**, *30*, 2445–2455. [CrossRef]

46. Romano, G.; Maresca, L.; Riccio, M.; d'Alessandro, V.; Breglio, G.; Irace, A.; Fayyaz, A.; Castellazzi, A. Short-circuit failure mechanism of SiC power MOSFETs. In Proceedings of the IEEE International Symposium on Power Semiconductor Devices and IC's (ISPSD), Hong Kong, China, 10–14 May 2015; pp. 345–348.

47. Cavallaro, D.; Pulvirenti, M.; Zanetti, E.; Saggio, M. Capability of SiC MOSFETs under short-circuit tests and development of a thermal model by finite element analysis. *Mater. Sci. Forum* **2019**, *963*, 788–791. [CrossRef]

Article

Influence of the Size and Shape of Magnetic Core on Thermal Parameters of the Inductor

Kalina Detka and Krzysztof Górecki *

Department of Marine Electronics, Gdynia Maritime University, Morska 83, 81-225 Gdynia, Poland;
k.detka@we.umg.edu.pl
* Correspondence: k.gorecki@we.umg.edu.pl

Received: 12 June 2020; Accepted: 24 July 2020; Published: 27 July 2020

Abstract: In this paper, a new thermal model of the inductor is proposed. This model takes into account self-heating in the core and in the winding, and mutual thermal couplings between the mentioned components of the inductor. The form of the elaborated thermal model is presented. In this model, the influence of power dissipated in the core and in the winding of the inductor on the efficiency of heat removal is taken into account. Correctness of the model is verified experimentally for inductors containing ferrite cores of different shapes and dimensions. The good agreement between the results of calculations and measurements is obtained. On the basis of the obtained findings, the influence of volume and the shape of the core on thermal resistances and thermal capacitances occurring in this model is discussed.

Keywords: inductors; ferromagnetic cores; thermal model; transient thermal impedance; thermal resistance; self-heating

1. Introduction

Inductors are important components of switch-mode power converters. These components are used to store electrical energy [1–5]. Properties of these components indeed depend on physical phenomena occurring in the winding and in the ferromagnetic core contained in the inductor [1,6–10]. At present, producers of ferromagnetic cores offer many types of cores made of different ferromagnetic materials. However, the most frequently used material to construct inductor cores are ferrites produced as a result of dwighting powdered metal oxides. Ferrites are characterised by high hardness, high resistivity, and low losses of eddy currents [5,9–14].

When an inductor operates in switched mode power converters an increase in power losses in this component is observed, which is the effect of current flowing through the winding of the inductor and remagnetisation of its core [1,5–16]. Power losses in the inductor are converted into heat. Heat generated in components of the inductor causes an increase in temperature of both the core and the winding above ambient temperature as a result of self-heating phenomena and mutual thermal couplings between the core and the winding of the inductor [8,11,15,17–19].

Temperature strongly influences the properties of electronic components, especially their reliability [20–22]. Therefore, it is essential to know the value of internal temperature of any element in the anticipated conditions of its operation. To calculate this temperature at well-known waveforms of power lost in the element, thermal models are used [23–26].

Thermal models presented in the literature have a character of microscopic models [6,7,27,28], dedicated to calculate distribution of temperature in the electronic component or macroscopic (compact) models [8,29–31], making it possible to calculate one value of internal temperature of the whole electronic component. Microscopic models, due to a high level of complexity, are not frequently used to analyse electronic circuits and they are used only to analyse thermal properties of single electronic components.

Typically compact thermal models are used. These models are often presented in the form of a network analogue [23,24,29,30,32]. Such an analogue usually consists of the current source, representing power dissipated in the modelled component and the RC network representing transient thermal impedance $Z_{th}(t)$, characterising the ability of a component to remove heat generated in this component [23–26]. Voltage on the current source corresponds to an excess of internal temperature of this component above ambient temperature [23–26].

The compact thermal model takes into account simultaneously all mechanisms of heat removal to the surroundings, i.e., conduction, convention, and radiation [24,30]. Transient thermal impedance occurring in the compact thermal model is typically described with the use of dependence of the form [24,29,30].

$$Z_{th}(t) = R_{th} \cdot \left[1 - \sum_{i=1}^{N} a_i \cdot \exp\left(-\frac{t}{\tau_{thi}}\right) \right] \tag{1}$$

where R_{th} is thermal resistance and N the number of thermal time constants τ_{thi} corresponding to coefficients a_i. At the steady-state the value of transient thermal impedance is equal to thermal resistance R_{th}.

In the literature many compact thermal models of inductors and transformers are described [6,8,15,17,33]. However, these models are highly simplified. Some of them [6,8] do not even take into account differences in temperature between the core and the winding. Thermal models of inductors taking into account nonlinearity of phenomena responsible for heat transfer are not known to the authors. Such nonlinearity is observed, among other things, in thermal models of semiconductor devices [30,34,35] or in the results of measurements of thermal properties of transformers shown among others in the papers [36,37]. Nowadays, there are no compact thermal models of inductors, which take into account influence of nonlinearity of thermal phenomena as well as the shape and the size of the core on thermal parameters of inductors.

The aim of this paper is to examine influence of the size and the shape of the ferromagnetic core and power losses on parameters of a thermal model of the inductor. In Section 2, a new nonlinear thermal model of the inductor is discussed. Section 3 describes a manner of estimation of parameters of the new model. The obtained results of calculations and measurements illustrating the usefulness of the proposed model are shown and discussed in Section 4.

2. New Nonlinear Thermal Model of Inductors

A nonlinear thermal model of inductors presented in this section was developed by the authors. It belongs to a group of compact thermal models of electronic components. Let us assume that these electronic components contain more than one heat source. Such models were described in the literature for IGBT (insulated gate bipolar transistor) modules [23] and LED (light emitting diode) modules [38] in which self-heating phenomena and mutual thermal couplings also occur.

The presented nonlinear thermal model of the inductor takes into account the fact that both the core temperature T_C and the winding temperature T_W depend on the value of ambient temperature T_a and a temperature excess, which is a result of a self-heating phenomenon in each component of the inductor and mutual thermal couplings between these components. The temperature excess of the inductor component caused by a self-heating phenomenon depends on the transient thermal impedance of the core $Z_{thC}(t)$ and transient thermal impedance of the winding $Z_{thW}(t)$. Thermal couplings between the core and the winding are characterised by mutual transient thermal impedance between the core and the winding $Z_{thCW}(t)$. Nonlinearity of phenomena responsible for transporting heat generated in the inductor to the surroundings is also taken into account.

The new nonlinear thermal model of the inductor is dedicated to the SPICE (simulation program with integrated circuits emphasis) program and has the network form shown in Figure 1.

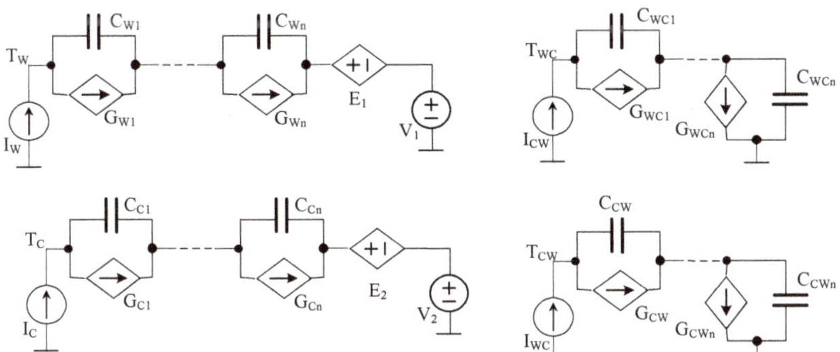

Figure 1. Nonlinear compact thermal model of the inductor.

This model consists of four subcircuits. Two of them, visible on the left side of Figure 1, allow for the calculation of the winding temperature T_w and the core temperature T_C. According to the rules of formulating a thermal model of electronic devices the Foster network is used [39]. In the mentioned network, current sources which describe the time dependence of power dissipated in the electronic element are used. Current sources I_W and I_C correspond to powers dissipated in the winding and in the core. Self-transient thermal impedances of the winding $Z_{thW}(t)$ and the core $Z_{thC}(t)$ are modelled using C_{Wi} and C_{Ci} capacitors and controlled current sources G_{Wi} and G_{Ci}. Voltages on these circuits correspond to a temperature excess caused by a self-heating phenomenon occurring in the winding and in the core. Influence of mutual thermal couplings between the core and the winding is modelled using the controlled voltage sources E_1 and E_2. Voltages on these sources correspond to the values of temperature excesses T_{WC} and T_{CW}. In contrast, voltage sources V_1 and V_2 model ambient temperature.

The other two subcircuits allow calculating excesses of the core temperature T_{CW} and the winding temperature T_{WC} caused by thermal coupling between the inductor components. In these subcircuits, current sources represent power dissipated in the winding I_{WC} and in the core I_{CW}. The networks connected to these sources model mutual transient thermal impedance between the core and the winding $Z_{thCW}(t)$. All transient thermal impedances occurring in the presented model are described by Equation (1).

As shown in the papers [23,35,36,40], waveforms of transient thermal impedance of electronic components depend on power or internal temperature of the considered electronic component. From the papers [23,36,38] and from measurements performed by the authors, it results that influence of power dissipated in the modelled component practically does not influence heat capacitances of this component, while influence of power dissipated in the considered component on thermal resistance could be significant. Therefore, in the presented nonlinear thermal model of the inductor, constant values of thermal capacitances are used. In contrast, thermal resistances occurring in the considered model depend on power generated in the inductor. Empirical dependence of thermal resistance R_{th} on the dissipated power p is proposed. This dependence is expressed by the empirical Equation:

$$R_{th} = R_{th0} + R_{th1} \cdot \exp\left(\frac{-p}{b}\right) \tag{2}$$

where is power dissipated in a heating component (core or winding), R_{th0}, R_{th1} and b are the model parameters. Of course, for each of the three transient thermal impedances (1), there is a different set of parameter values describing thermal capacitances and thermal resistances.

Changes of individual components of thermal resistance are modelled using the controlled current sources G_i according to the Equation [33]:

$$G_i = \frac{V_{Gi}}{a_i \cdot R_{th}} \tag{3}$$

where V_{Gi} is voltage on source G_i, R_{th} thermal resistance calculated using the Equation (2), and a_i is the coefficient in the Equation (1) which corresponds to i-th thermal time constants τ_{thi}.

Values of parameters occurring in Equations (1) and (2) were determined using the concept of local estimation [41,42] based on the measured waveforms of transient thermal impedances occurring in the considered thermal model of the inductor.

3. Method of Model Parameters Estimation

Values of parameters of the thermal model described in Section 2 can be determined as a result of realisation of a series of measurements and calculations. These measurements are performed in the measuring set-up shown in Figure 2.

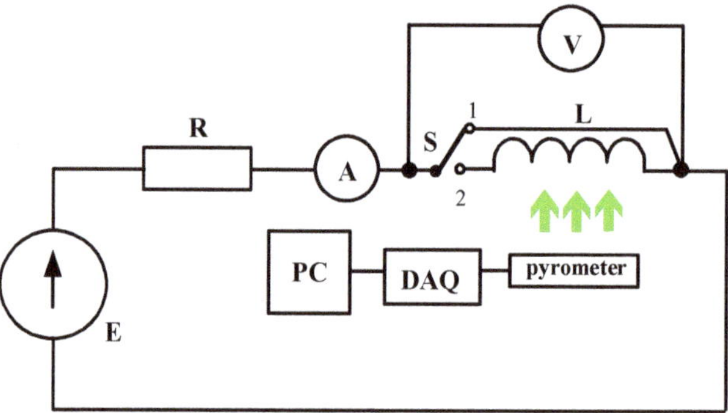

Figure 2. Set-up to measure thermal parameters of the inductor.

The considered set-up contains a voltage source E, a resistor R limiting the value of current, an ammeter, a voltmeter, the examined inductor L, a pyrometer, the acquisition data system DAQ, and a PC. To measure the value of DC current and DC voltage multimeters of the type UNIT UT-803 were used. The uncertainty of the measurements of DC voltage is ±0.025%, of the DC current ±0.1%, and of temperature measured by the pyrometer PT-3S is equal to ±3% [43].

In this set-up transient thermal impedances of the core $Z_{thC}(t)$ and the winding $Z_{thW}(t)$, and also mutual transient thermal impedance between the core and the winding $Z_{thCW}(t)$ are measured. These measurements are realised with the use of the indirect method and the following definitional Equations.

$$Z_{thC}(t) = \frac{T_C(t) - T_a}{p_C} \tag{4}$$

$$Z_{thW}(t) = \frac{T_W(t) - T_a}{p_W} \tag{5}$$

$$Z_{thWC}(t) = \frac{T_C(t) - T_a}{p_W} \tag{6}$$

where p_C denotes power dissipated in the inductor core, whereas p_W refers to power dissipated in the winding. In Equations (4)–(6), the temperatures of the core and the winding and powers dissipated in the core and in the winding also appear.

In the considered measuring set-up power in the shape of a jump is dissipated during the flow of current through one of two components of the inductor depending on the position of switch S. In the position 1 of the switch, power is dissipated in the core, and in the position 2 of this switch power is dissipated in the winding. The value of current flowing through components of the inductor is regulated by means of voltage source E and resistor R. The temperature of the core is registered by means of the pyrometer PT-3S [43] configured to work in the continuous operation, as well as the card of data acquisition and a computer. In turn, temperature of the winding is measured indirectly on the basis of measurements of the winding resistance.

The value of voltage and current flowing through the core or the winding is regulated over a wide range of the measured waveforms $Z_{thC}(t)$, $Z_{thW}(t)$ and $Z_{thCW}(t)$ for different values of power p_C and p_W. Basing on the registered waveforms of the mentioned transient thermal impedances of the inductor, values of parameters R_{th}, a_i, τ_{thi} occurring in the Equation (1) for every applied value of power p_W and p_C are estimated using the program ESTYM [42]. For every transient thermal impedance of the modelled inductor at the highest applied values of the dissipated power, average values of parameters a_i and thermal capacitances occurring in the proposed thermal model are calculated on the basis of the Equation:

$$C_i = \frac{\tau_{thi}}{a_i \cdot R_{th}} \tag{7}$$

Based on the measured dependences of thermal resistance on power $R_{thC}(p_C)$, $R_{thW}(p_W)$, and $R_{thCW}(p_W)$, values of parameters R_{th0}, R_{th1}, and b occurring in Equation (2) are estimated with the method of local estimation [39,40] for every considered dependence separately.

4. Results

In order to analyse influence of the dimensions and the shape of the core on parameters of a non-linear thermal model of the inductor, investigations of the considered component operating in the set-up presented in Figure 2 were performed. Measurements were carried out for inductors with ferrite cores of different shapes and dimensions.

In subsection A the tested inductors are described, in subsection B presents the estimated values of model parameters, and in subsection C, the obtained results of calculations and measurements are presented. On the basis of the obtained results dependences of influence of the mentioned factors on the value of parameters of the presented nonlinear thermal model of the inductor are discussed.

4.1. Tested Inductors

Inductors containing cup and toroidal cores made of ferrite material F-867 [44] were used for investigations. On each core, eight turns of copper wire in the enamel of the diameter 1 mm were wound. The examined inductors with cup cores of different dimensions were mounted on the printed circuit board, which was situated vertically during measurements. The inductor with the ring core was also arranged vertically. In Figure 3, the dimensions of the examined inductor cores are shown.

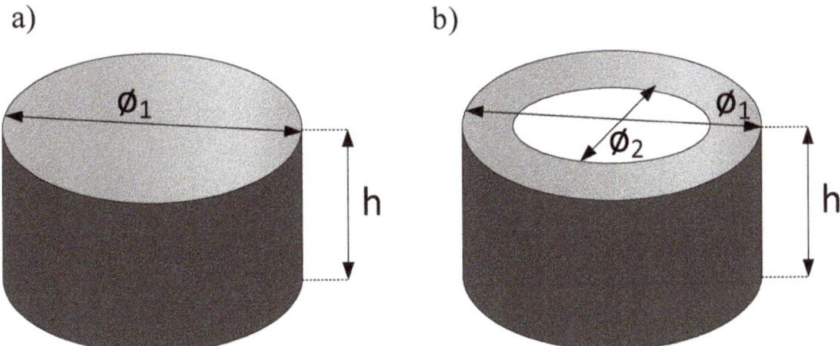

Figure 3. Dimensions of the investigated (**a**) cup core and (**b**) toroidal core.

In Figure 4, the examined inductors with cup cores installed on the printed circuit board are shown, and in Figure 5 the examined inductor with the ring core is presented. During measurements ambient temperature was monitored and its value fluctuated between 21.9 °C and 23 °C. In both the figures, cables mounted to the ferromagnetic core are visible. These cables are indispensable to enable the current flow through the ferromagnetic core while heating this core. In Table 1 values of selected parameters of material F867 are collected, whereas in Table 2, values of geometrical parameters of the considered inductor cores are given. The inductors with the core whose parameters are collected in Table 2 are shown in Figures 4 and 5.

 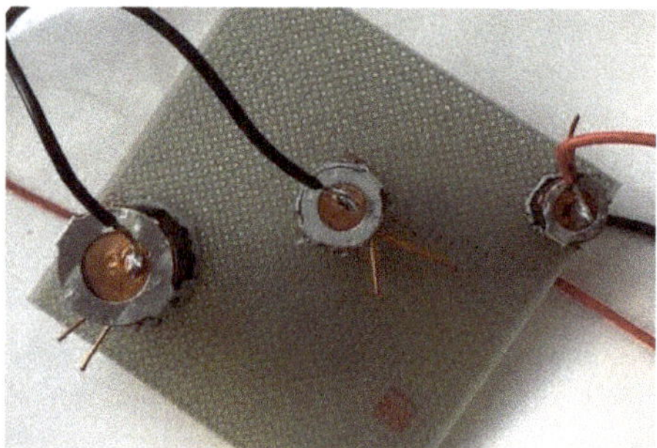

Figure 4. Tested inductors containing cup cores of different dimensions.

Figure 5. Tested inductors containing toroidal cores.

Table 1. Values of selected parameters of material F867 [43].

Parameter	B_{sat} (T)	B_R (T)	H_C (A/m)	μ_i	P_v (mW/cm^3)
Value (25 °C)	0.6	0.15	40	2400	129
Value (100 °C)	0.4	0.12	40	3900	70

Table 2. Geometrical parameters of the tested inductor cores.

Cup Core					
Dimensions (mm)		l_e (mm)	A_e (mm^2)	V_e (mm^3)	
$\varnothing_1$	h				
14	8	20.6	23.1	485	
18	11	25.9	43	1120	
26	16	37.2	93	3460	

Toroidal Core					
Dimensions (mm)			l_e (mm)	A_e (mm^2)	V_e (mm^3)
$\varnothing_1$	$\varnothing_2$	h			
16	9.5	6.5	40.03	21.12	845.71
20	10	10	35.7	150	2355
31	19	13	78.5	78	6123
40	24	15	98.56	448	12861

As shown in Table 1, saturation flux density B_{sat} decreases with a temperature increase from 0.6 T to 0.4 T, remanence flux density B_R does not exceed 0.15 T, coercion force H_C amounts to 40 A/m, initial permeability μ_i strongly depend on temperature and increases from 2400 to 3900. Power losses per unit of volume P_V decrease from 129 mW/cm^3 to 70 mW/cm^3 in the considered changes of temperature. In further part of this paper the considered cup cores will be denoted as: small cup core, medium cup core and big cup core, respectively. In turn, toroidal cores will be denoted as: toroidal core 16, toroidal core 20, toroidal core 30 and toroidal core 40, respectively.

Table 2 shows that the used cores are characterised by different values of such geometrical parameters as: magnetic path length l_e, cross-section area A_e and volume V_e. For example toroidal core 16 has similar value of l_e parameter to big cup core l_e parameter and similar value of A_e parameter to small cup core A_e parameter. Further, the toroidal core 20 has similar value of l_e parameter to big cup core.

4.2. Parameters Values of a New Model

Using the measurement set-up (Figure 2), measurements of transient thermal impedances were performed. They were carried out for all the considered inductors at different values of power dissipated in the cores and windings. Based on the obtained measurement results, values of parameters of the nonlinear thermal model were determined for each considered inductor with the use of the method described in Section 2. For example, values of parameters of this model for inductors with the medium cup core and with the toroidal core of the external diameter equal to 16 mm are presented in Table 3.

Table 3. Values of parameters of the nonlinear thermal model of inductors with the medium cup core and with the toroidal core.

Parameter	R_{th0} (K/W)	R_{th1} (K/W)	b (W)	a_1	a_2	C_{th1} (J/K)	C_{th2} (J/K)
		Inductor with the Medium Cup Core					
$Z_{thW}(t)$	25	11	2	0.403	0.597	2.403	8.07
$Z_{thC}(t)$	19	15	2	0.449	0.551	10.694	23.693
$Z_{thWC}(t)$	15	12	1.4	0.937	0.063	13.99	219.56
		Inductor with the Toroidal Core 16					
$Z_{thW}(t)$	25	20	5.3	0.26	0.74	0.516	4.958
$Z_{thC}(t)$	20	15	4	0.994	0.006	5.132	151.47
$Z_{thWC}(t)$	15	8	19	0.906	0.094	7.605	53.85

As can be observed, for both the inductors, the same number of thermal time constants, related to the coefficients a_i, which describe particular transient thermal impedances, is obtained. Values of thermal capacitances characterising thermal properties of the core are higher than those capacitances characterising the winding properties. Values of parameters R_{th0} appearing in the description of individual transient thermal impedances are similar for both the considered inductors. In contrast, even ten-fold differences are observed between values of b parameter describing the considered transient thermal impedances.

Table 4 compares the values of parameters of transient thermal impedance of the core $Z_{thC}(t)$ obtained for the inductor with cup cores of different sizes.

Table 4. Values of parameters describing $Z_{thC}(t)$ of inductors with cup cores.

Parameter	R_{th0} (K/W)	R_{th1} (K/W)	b (W)	a_1	a_2	C_{th1} (J/K)	C_{th2} (J/K)
Small core	38	33	0.5	0.35	0.65	1.114	6.503
Medium core	20	15	4	0.449	0.551	11.601	25.172
Big core	16	10	1	0.023	0.977	27.172	29.228

As it is visible, an increase in the dimensions of the cup core causes a decrease in the value of parameters R_{th0} and R_{th1}, whereas parameter b achieves maximum value for medium cup core. Also an increase in thermal capacitances values with core size is observed.

Table 5 compares values of parameters of transient thermal impedance of the winding $Z_{thW}(t)$ obtained for the inductor with cup cores of different sizes.

Table 5. Values of parameters describing $Z_{thW}(t)$ of inductors with cup cores.

Parameter	R_{th0} (K/W)	R_{th1} (K/W)	b (W)	a_1	a_2	C_{th1} (J/K)	C_{th2} (J/K)
Medium core	25	11	2	0.319	0.681	2.498	9.620
Big core	7	50	2	0.39	0.61	3.164	19.043

As can be seen, an increase in the dimensions of the cup core causes a decrease in the value of parameters R_{th0} and R_{th1}, whereas parameter b have the same value for both considered cup cores. Additionally, increase in dimensions of the cup core causes a visible increase in the value of thermal capacitance.

Table 6 collects the values of parameters of transient thermal impedance of the core $Z_{thC}(t)$ obtained for the inductor with toroidal cores of different dimensions.

Table 6. Values of parameters describing $Z_{thC}(t)$ of inductors with toroidal cores of different dimensions.

Parameter	R_{th0} (K/W)	R_{th1} (K/W)	b (W)	a_1	a_2	C_{th1} (J/K)	C_{th2} (J/K)
toroidal core 16	24	13	2	0.994	0.006	4.79	27.07
toroidal core 20	19.6	5.5	1	0.874	0.126	17.530	28.43
toroidal core 30	11.8	7.9	3.3	0.98	0.02	27.265	233.12
toroidal core 40	11	6	1	0.984	0.016	63.356	550.80

As shown, an increase in the dimensions of the toroidal core causes an increase in the value of parameters R_{th0} and thermal capacitances.

Table 7 collects values of parameters of transient thermal impedance of the winding $Z_{thW}(t)$ obtained for the inductor with toroidal core of different dimensions.

Table 7. Values of parameters describing $Z_{thw}(t)$ of inductors with toroidal core.

Parameter	R_{th0} (K/W)	R_{th1} (K/W)	b (W)	a_1	a_2	C_{th1} (J/K)	C_{th2} (J/K)
toroidal core 16	25	20	5.3	0.26	0.74	0.516	4.958
toroidal core 20	19.4	7	1.2	0.179	0.821	1.878	18.976
toroidal core 30	11.8	7.9	3.3	0.192	0.808	5.650	27.265
toroidal core 40	8	24	1.3	0.275	0.725	4.383	44.125

As can be observed, an increase in the toroidal core dimensions causes an increase in thermal capacitance and a decrease in the parameter R_{th0}. For example, parameter R_{th0} decreases even triple when the diameter of the core increases 2.5 times.

4.3. Results of Measurements and Calculations

In order to verify the usefulness of the thermal model of inductors proposed in Section 2, some measurements and computation were performed. In computations the nonlinear thermal model of the inductor was used. Results of these measurements and computations are shown in Figures 5–17.

In these figures, the results obtained for particular inductors were marked using the following markers and colour rules: an inductor with 14×8 mm dimensions of the cup core is marked as the small cup core (blue), an inductor with 18×11 mm dimensions of the cup core, the medium cup core (green), and an inductor with 26×16 mm dimensions of the cup core is marked as the big cup core (red). An inductor containing toroidal core with 16 mm diameter of the core is marked as a toroidal core

16 (violet), an inductor with 20 mm diameter of the core is marked toroidal core 20 (green), an inductor with 30 mm diameter of the core is marked toroidal core 30 (yellow) and an inductor with 40 mm diameter of the core is marked toroidal core 40 (blue). Additionally, it is worth remembering that the volume of toroidal core with 16 mm diameter corresponding to the volume of medium cup core. In all the figures, lines denote the results of calculations, whereas points refer to the results of measurements.

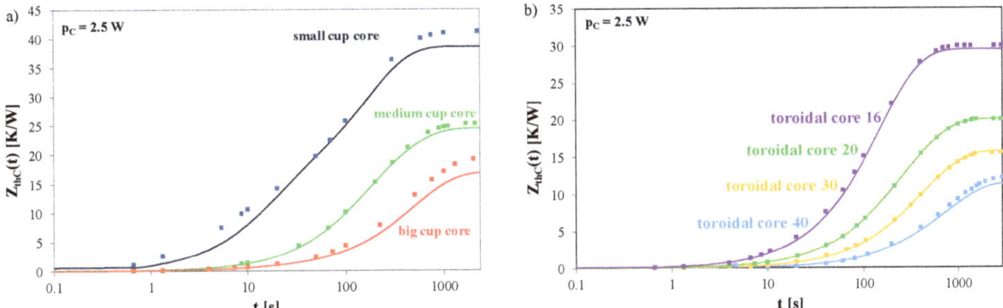

Figure 6. Measured and calculated waveforms of transient thermal impedance of the core for inductors with (**a**) cup and (**b**) toroidal cores of different dimensions.

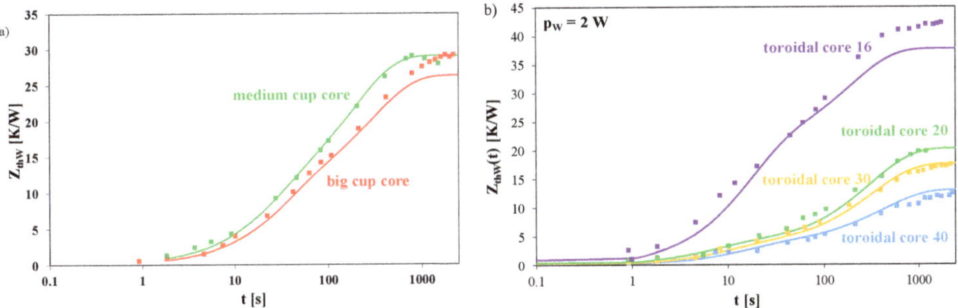

Figure 7. Measured and calculated waveforms of transient thermal impedance of the winding for inductors with (**a**) cup and (**b**) toroidal cores.

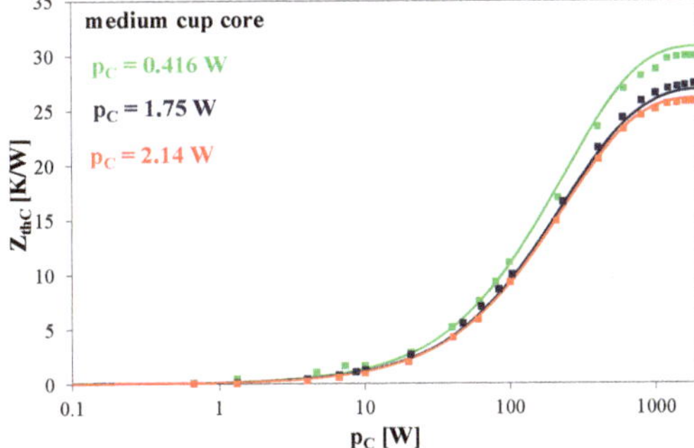

Figure 8. Measured and calculated waveforms of transient thermal impedance of the medium cup core at selected values of dissipated power.

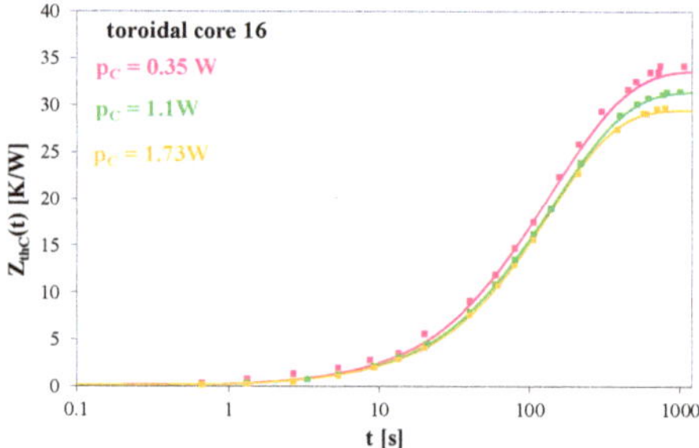

Figure 9. Measured and calculated waveforms of transient thermal impedances of the inductor containing the toroidal core 16.

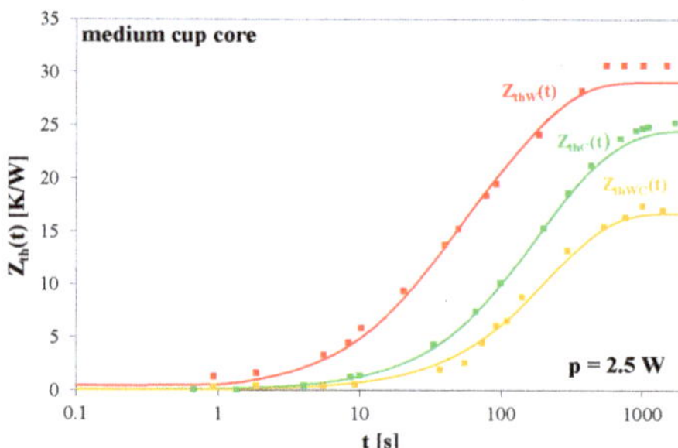

Figure 10. Measured and calculated waveforms of transient thermal impedances of the inductor containing the medium cup core.

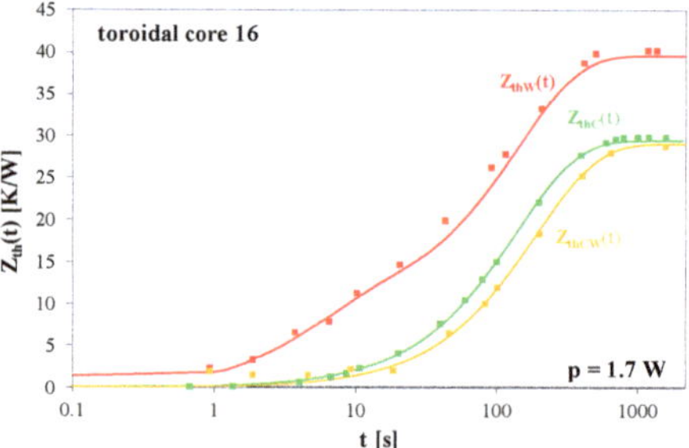

Figure 11. Measured and calculated waveforms of transient thermal impedances of the inductor containing the toroidal core 16.

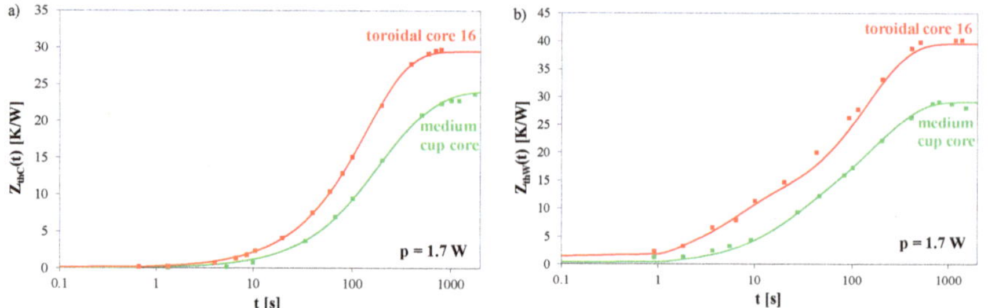

Figure 12. Measured and calculated waveforms of transient thermal impedances of the core (**a**) and of the winding (**b**) for inductors containing each considered core.

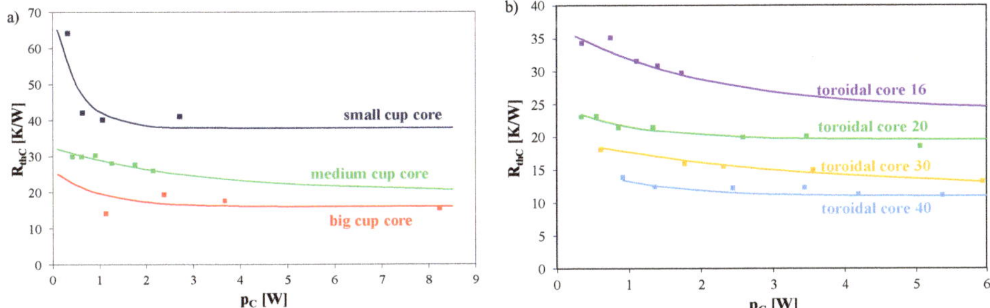

Figure 13. Measured and calculated dependences of thermal resistance R_{thC} of inductors with (**a**) cup cores and (**b**) toroidal core on dissipated power in the core.

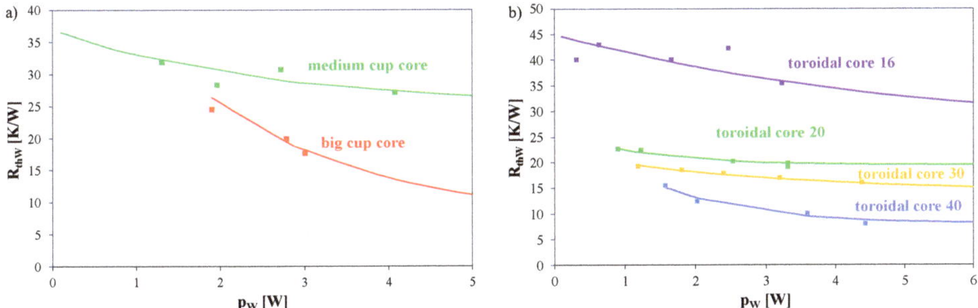

Figure 14. Measured and calculated dependences of thermal resistance R_{thW} of inductors winding with (**a**) cup cores and (**b**) toroidal core on dissipated power in the winding.

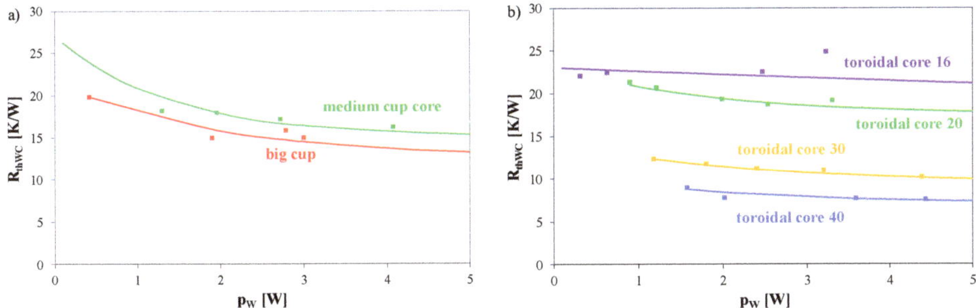

Figure 15. Measured and calculated dependences of thermal resistance R_{thW} of inductors with (**a**) cup cores and (**b**) toroidal core on dissipated power in the winding.

At first, measured and calculated waveforms of transient thermal impedances occurring in the proposed thermal model of an inductor are presented. Next, dependences illustrating an influence of dissipated power in components of the tested inductors on thermal resistances occurring in the considered model are shown and discussed. Finally, an analytical description of the dependences of thermal resistances and capacitances on the effective volume of the core contained in the tested inductors are proposed and experimentally verified for these inductors.

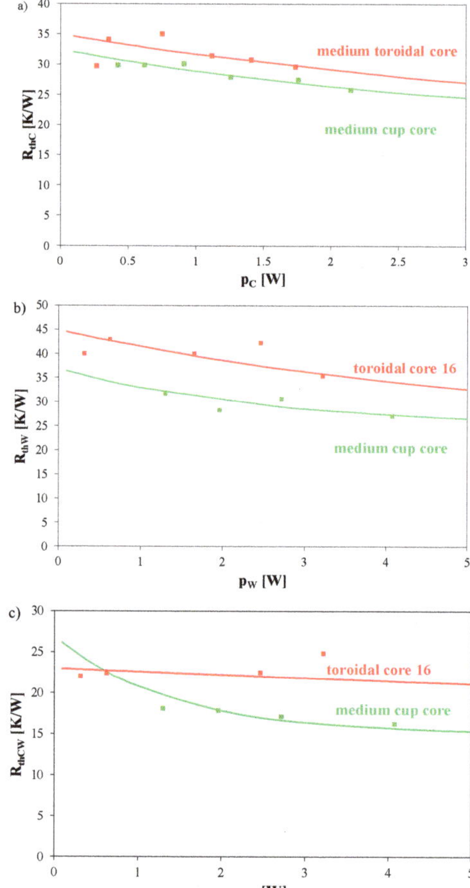

Figure 16. Measured and calculated dependences of thermal resistances (**a**) R_{thC}, (**b**) R_{thW} and (**c**) R_{thCW} of inductors with the medium cup core and the toroidal core on power dissipated in the core.

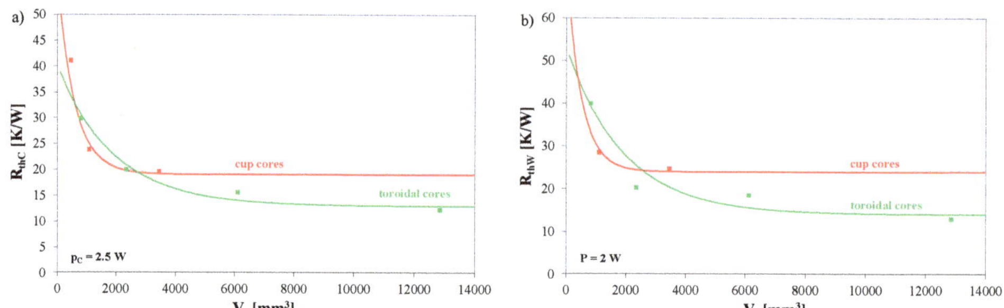

Figure 17. Measured and calculated dependences of thermal resistance of the core R_{thC} (**a**) and thermal resistance of the winding R_{thW} (**b**) on the effective volume of the core for inductors with cup cores and toroidal cores.

Figure 6 presents the calculated and measured waveforms of transient thermal impedance of the core of the considered inductors containing cup cores (Figure 6a) and toroidal core (Figure 6b) obtained at dissipation in the core the power of the amplitude equal to 2.5 W.

As can be seen, from the obtained waveforms of transient thermal impedance of the core, for the smallest volume of the core, the value of $Z_{thC}(t)$ at the steady state is more than twice higher than the value of $Z_{thC}(t)$ at the steady state for the biggest volume of the core (big cup core) and over 60% higher than the value $Z_{thC}(t)$ at the steady state for the core of medium volume—the medium cup core. It can be concluded from this relation that the ability to remove heat characterised by thermal resistance of the core R_{thC} decreases with an increase in the core size. This is due to an increase in the surface area, at which convection heat transfer rate can occur. On the other hand, the time needed to reach the thermally steady state for the big core of the inductor is more than twice longer for the small inductor core. This means that thermal capacitance of the core increases with its size. In the case of the inductor with the toroidal core with the 16 mm diameter the value of $Z_{thC}(t)$ at the steady state is more than twice higher than the value of $Z_{thC}(t)$ at the steady state for the toroidal core with the 40 mm diameter and about 50% higher than the value of $Z_{thC}(t)$ at the steady state for the toroidal core with the 20 mm diameter. It is also worth noticing that the good agreement between the results of measurements and the results of calculations was obtained. For the toroidal core, the maximum error of calculations does not exceed 5% and for the cup core it is smaller than 8%.

Figure 7 presents the calculated and measured waveforms of transient thermal impedance of the winding of the considered inductors containing cup cores (Figure 7a) and toroidal cores (Figure 7b) obtained at power dissipated in the core equal to 2 W.

As can be seen, from the obtained waveforms of transient thermal impedance of the winding of the inductor with cup cores (Figure 7a), for the medium volume of the core, the value of $Z_{thW}(t)$ at the steady state is the same as the value of $Z_{thW}(t)$ at the steady state for the big cup core. In the case of the inductor with the toroidal core with the 16 mm diameter the value of $Z_{thW}(t)$ at the steady state for the considered inductor is five times higher than for the same inductor with the core with the 40 mm diameter. Additionally, differences between the results of calculations and measurements do not exceed 11% for the inductor with the cup core and 13% for the inductor with the toroidal core.

Figures 8 and 9 show the waveforms of thermal transient impedance of the medium cup core (Figure 8) and of the toroidal core (Figure 9) for selected values of power dissipated in the core.

As can be seen in Figure 8, due to changes in power dissipated in the cup core, values of $Z_{thC}(t)$ at the steady state change by not more than 20%. An increase in the value of power causes a decrease of $Z_{thC}(t)$ value. It is observed that power does not influence time indispensable to achieve the steady state of $Z_{thC}(t)$ waveform. This means that thermal capacitance is practically independent of power dissipated in the core.

However, an increase in the value of power dissipated in the toroidal core (Figure 9) causes a decrease in the value of $Z_{thC}(t)$ at the steady state. These changes reach almost 15%. At the same time, it can be seen that the value of power dissipated in the core practically does not influence time, in which the waveform of $Z_{thC}(t)$ achieves the steady state. As can be seen, differences between the results of calculations and measurements do not exceed 3%.

Figures 10 and 11 present the measured and calculated waveforms of transient thermal impedances $Z_{thW}(t)$, $Z_{thC}(t)$, and $Z_{thCW}(t)$ for the inductor containing the medium cup core (Figure 10) at power dissipated in the core and in the winding equal to 2.5 W, and for the toroidal core, 16 (Figure 11) at power dissipated in the core $p_C = 1.7$ W.

It can be seen that, at the steady state, values of $Z_{thW}(t)$ are up to 50% higher than $Z_{thC}(t)$. Time necessary to obtain the steady state is the shortest for waveform $Z_{thW}(t)$ and the longest for $Z_{thC}(t)$. Differences in the values of these times reach 20%, and are a result, among others, of differences in the mass of the core and windings. Also, the good agreement between the results of measurements and calculations was obtained. The maximum deviation does not exceed 8%.

Similarly to the cup core, the highest values are obtained for $Z_{thW}(t)$ (Figure 11). They are even 30% higher than the value of $Z_{thC}(t)$. Waveforms of $Z_{thC}(t)$ and $Z_{thCW}(t)$ differ from each other by not more than 5%. These differences are due to construction of the inductor, which causes that the winding is directly cooled by the air surrounding the inductor, and due also to the fact that the core is surrounded by the winding, in contrast to the cup core, which is not. Also the good agreement between the results of measurements and calculations is achieved and the differences between them do not exceed 3%.

Figure 12 presents a comparison of waveforms of transient thermal impedances of the core $Z_{thC}(t)$ and the winding $Z_{thW}(t)$ for the inductor containing the toroidal core and the cup core at power dissipated equal to 1.7 W.

The presented comparison shows that values of transient thermal impedances $Z_{thC}(t)$ and $Z_{thW}(t)$ are about 30% higher for the inductor with the toroidal core. The setting time for $Z_{thC}(t)$ waveform is longer for the inductor with the cup core, while the setting time for $Z_{thW}(t)$ waveform is practically the same. In this case, the error of calculations does not exceed 3%.

Figures 13–15 shows the calculated with the use of the Equation (2) and measured dependences of thermal resistances R_{thC} (Figure 13), R_{thW} (Figure 14), R_{thWC} (Figure 15) occurring in the thermal model of the considered inductors with cup cores and inductors with toroidal cores on power dissipated in these components of the inductor.

As can be seen in Figure 13, dependence $R_{thC}(p_C)$ is a decreasing function for both the inductors with the cup cores (Figure 13a) and with the toroidal core (Figure 13b). It is also visible that as the core size increases, thermal resistance values decrease. The biggest differences in the values of this parameter for the considered cores can be seen in the range of low values of power p_C. Differences between the calculation results and the measurement results do not exceed a dozen percent for the inductor with the cup core and do not exceed 8% for the inductor with the toroidal core.

Figure 14 presents the calculated and measured dependences of thermal resistance R_{thW} of the winding of the considered inductors with cup cores (Figure 14a) and inductors with toroidal cores (Figure 14b) on power dissipated in these cores. Due to the limited size of the small core, it was impossible to wind eight turns of copper wire in enamel with a diameter of 1 mm on inductor cores, so in the following results, a comparison between the big and medium cup cores only are presented in Figure 13a.

The characteristics presented in Figure 14 have a similar shape as the characteristics presented in Figure 13. An increase in the core volume causes a decrease in thermal resistance of the winding. The differences between the results of measurements and calculations for inductors with the cup core do not exceed 10% for all the considered inductors with cup and toroidal cores. It is also worth noticing that thermal resistance of the inductor containing the toroidal core with the 16 mm diameter is higher by even 10 K/W than thermal resistance of the inductor containing the medium cup core. Additionally, the differences between the results of measurements and calculations do not exceed 11% for the inductor with the cup core and 15.5% for the inductor with the toroidal core.

Figure 15 presents the calculated and measured dependences of mutual thermal resistance between winding and core R_{thWC} of the inductors with cup cores (Figure 15a) and inductors with toroidal cores (Figure 15b) on power dissipated in these cores.

As can be seen, an increase of the core size causes a decrease of thermal resistance of the considered inductors with cup and toroidal cores. In the case of the inductor with the toroidal core an increase in diameter from 16 to 40 mm causes a more than double decrease of thermal resistance, whereas an increase of the diameter from 18 to 26 mm of the cup core causes a decrease in thermal resistance by about 15%. Differences between the calculations and measurements results do not exceed a dozen per cent for the inductor with the cup core and they do not exceed 13% for the inductor with the toroidal core.

Figure 16 presents the measured and calculated dependences of thermal resistances R_{thC}, R_{thW} and R_{thCW} occurring in the nonlinear thermal model of the inductor on power dissipated in the core

(Figure 16a) and in the winding (Figure 16b) for inductors with the medium cup core and the toroidal core 16. As mentioned in Section 3, both the considered cores have similar volume.

As can be seen, dependences of all the thermal resistances on power dissipated in the core are decreasing functions. Values of the considered parameters for the toroidal core are higher than the values obtained for the cup core. The highest values were obtained for thermal resistance of the winding R_{thW}, and the lowest values for mutual thermal resistance between the core and the winding R_{thCW}. Values of these parameters differ between each other even twice. Due to the influence of changes in power values in the considered range, changes in individual thermal resistances up to 20% are observed. Also, the good agreement between the results of calculations and measurements was obtained. The differences between the results of calculations and measurements are less than 12%.

Analysing results of investigation presented above we formulated the analytic Equation describing an influence of core volume on thermal resistances existing in the thermal model of the inductor. The form of this Equation is as follows:

$$R_{th} = R_{thA} \cdot \left(1 + k_1 \cdot \exp\left(-\frac{V_e}{m_1}\right)\right) \tag{8}$$

where R_{thA} denotes border value of the thermal resistance at the volume of core tending to infinity, V_e is equivalent core volume, whereas m_1 and k_1 are model parameters characterising the slope of the dependence $R_{th}(V_e)$.

In the same way, an analytical description of the dependence of thermal capacitance on core volume was formulated. This dependence is given by following Equation:

$$C_{th} = C_{thA} \cdot (1 + k_3 \cdot V_e) \tag{9}$$

where C_{thA} denotes border value of thermal capacitance corresponding to zero value of volume V_e, whereas k_3 is volume coefficient of thermal capacitance.

Figure 17 presents the measured (points) and calculated (lines) using Equation (8) dependences of thermal resistance of the core R_{thC}, occurring in the nonlinear thermal model of the inductor on effective volume of the core for inductors with the cup cores (red colour) and the toroidal cores (green colours). Measurements and computations were performed at power dissipated in the core equal to 2 W.

As it is visible, for both the considered shapes of the core the dependence $R_{th}(V_e)$ is a decreasing function. Values of both thermal resistances for cup cores are smaller than for toroidal core in the range of low values of core volume, whereas in the range of high values of core volume these relation is opposite. It is worth noticing that in the considered range of change the core volume values of thermal resistance decreases over twice. For both the shapes of core, a good accuracy of modelling considered dependences are obtained.

Figure 18 illustrates an influence of core volume on selected thermal capacitances occurring in thermal model of tested inductors.

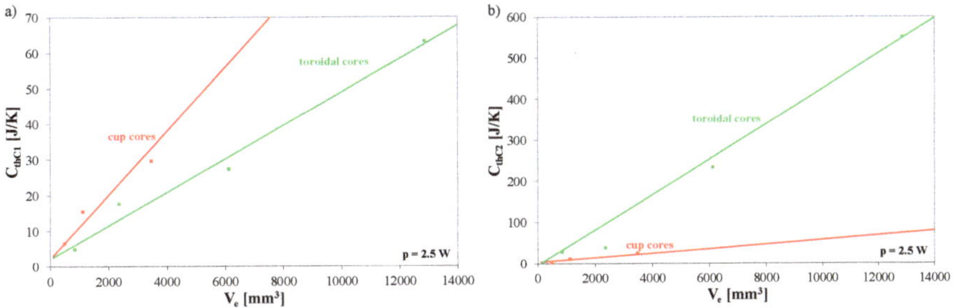

Figure 18. Measured and calculated dependences of selected thermal capacitance of the core C_{thC1} (**a**) and C_{thC2} (**b**) on the effective volume of the core for inductors with cup cores and toroidal cores.

As can be observed, the considered dependences are increasing functions. It is worth noticing that the thermal capacitance C_{thC1} of the tested inductors is smaller for inductors including toroidal cores, whereas the thermal capacitance C_{thC2} is smaller for inductors including cup cores. Also, the good agreement between the results of measurements and calculations was obtained. The differences between these results do not exceed 12% for both considered shapes of cores.

5. Conclusions

In the paper, a compact nonlinear thermal model of the inductor was proposed. This model makes it possible to calculate values of temperature of the core and the winding of the inductor taking into account occurrence of self-heating in every mentioned component of the inductor and mutual thermal couplings between the core and the winding. It also takes into account the influence of power dissipated in every component of the inductor on thermal resistance of the core and the winding and mutual thermal resistance between the core and the winding. A manner of calculating the value of parameters of this model was also proposed.

Correctness of the worked out model was verified for selected inductors containing ferrite cores made of the same ferrite material, but these cores were characterised by a different shape or by a different size. As a result of the comparison of the obtained results of calculations and measurements it was shown that the elaborated model is universal, i.e., it makes it possible to obtain the good agreement of these results over a wide range of changes of power dissipated in each component of the inductor at different shapes and dimensions of the core.

Comparing the findings obtained for different sizes of cup cores, it was observed that an increase in the dimensions of the core of the considered shape caused a decrease in the value of thermal resistance and extension of time indispensable to obtain the thermally steady state in the examined inductor. Taking into account the fact that a basic mechanism of removing heat generated in the core of the inductor is convection, it can be said that the value of thermal resistance of the core is a decreasing function of the surface of the cup core. Referring to the results of measurements shown in the paper [36] it can be stated that in the description of the considered dependence spatial orientation of the inductor should be also taken into account. In turn, the thermal capacitance of the core, deciding the time of settlement of the waveform of transient thermal impedance, depends on the volume of the ferrite core. Then, the thermal capacitances of the core can be described with an increasing function of the volume of the core.

The authors proposed analytical Equations describing dependences of thermal resistances and thermal capacitances of the core and the winding of the inductor on the volume of the core. Correctness of the formulated Equations was proved for both the considered shapes of cores and a good match between the results of measurements and calculations was obtained. The differences between the results of calculations and measurements do not exceed 15% maximum. The obtained results of calculations performed using the new thermal model of the inductor confirm usefulness of the formulated model.

A change in the shape of the core also influences waveforms of transient thermal impedances occurring in the new nonlinear compact thermal model of the inductor. At the similar volume of the core, greater even by 20% values of thermal resistance were obtained for the inductor with the cup core. The observed changes in the value of thermal resistances in the function of volume of the core are higher for inductors with cup cores than inductors with toroidal cores. From the thermal management point of view, it is more profitable to use toroidal cores than cup cores.

The obtained results of investigations make it possible to model the thermal properties of inductors in a simple way. The proposed thermal model of inductors can be used in power electronics applications. In the mentioned applications, properties of magnetic elements strongly influence watt-hour efficiency. Using the new model, the designers of power electronic circuits can calculate thermal parameters and temperature of every component of the designed inductor. They can also determine usefulness of selected inductors in the anticipated operating conditions of the designing step.

The results of investigations presented in this paper correspond to one ferromagnetic material only. In further investigations other ferromagnetic materials and other shapes of the cores will be analysed.

Author Contributions: Conceptualization, K.G.; methodology, K.G. and K.D.; measurements, K.D.; computations, K.G. and K.D.; resources, K.D.; writing—original draft preparation, K.G. and K.D.; writing—review and editing, K.G. and K.D.; visualization, K.G. and K.D.; supervision, K.G. All authors have read and agreed to the published version of the manuscript.

Funding: Project financed in the framework of the program by Ministry of Science and Higher Education called "Regionalna Inicjatywa Doskonałości" in the years 2019–2022, project number 006/RID/2018/19, the sum of financing 11,870,000 PLN.

Conflicts of Interest: The authors declare no conflict of interest.

Nomenclature

Symbol	Unit	Explanation
R_{th}	K/W	thermal resistance
N		number of thermal time constant
τ_{thi}	s	thermal time constants
a_i		coefficients whose sum has to be equal 1
$Z_{thC}(t)$	K/W	thermal impedance of the core
$Z_{thW}(t)$	K/W	thermal impedance of the winding
$Z_{thCW}(t)$	K/W	mutual thermal impedance between the core and the winding
p_C	W	power dissipated in the core
p_W	W	power dissipated in the winding
P_v	mW/cm^3	power losses per unit of volume in the core
T_a	°C	ambient temperature
T_W	°C	winding temperature
T_C	°C	core temperature
B_{sat}	T	saturation flux density
B_R	T	remanence flux density
H_C	A/m	coercion magnetic force
μ_i		initial permeability
l_e	mm	magnetic path length
V_e	mm^3	volume of the core
A_e	mm^2	cross-section area of the core
R_{thA}	K/W	value of thermal resistance at the volume of the core tending to infinity
m_1	mm^3	model parameters characterising the slope of dependence $R_{th}(V_e)$
k_1		model parameters characterising the slope of dependence $R_{th}(V_e)$
C_{thA}	J/K	value of thermal capacitance corresponding to zero value of volume V_e
k_3	mm^{-3}	volume coefficient of thermal capacitance

References

1. Barlik, R.; Nowak, M.; Grzejszczak, P.; Zdanowski, M. Estimation of power losses in a high-frequency planar transformer using a thermal camera. *Arch. Electr. Eng.* **2016**, *65*, 613–627. [CrossRef]
2. Rashid, M.H. *Power Electronic Handbook*; Academic Press: Cambridge, MA, USA, 2007.
3. Billings, K.; Morey, T. *Switch-Mode Power Supply Handbook*; McGraw-Hill: New York, NY, USA, 2011.
4. Erickson, R.; Maksimović, D. *Fundamentals of Power Electronics*; Kluwer Academic Publisher: Norwell, MA, USA, 2001.
5. Van den Bossche, A.; Valchev, V. *Inductor and Transformers for Power Electronic*; CRC Press: Boca Raton, FL, USA, 2005.
6. Andreu, D.; Boucher, J.; Maxim, A. New SPICE behavioural macromodelling method of magnetic components including the self-heating process. In Proceedings of the IEEE Annual Power Electronics Specialist Conference PESC, Charleston, SC, USA, 27 June–1 July 1999; Volume 2, pp. 735–740.
7. Allahbakhshi, M.; Akbari, A. An improved computational approach for thermal modeling of power transformers. *Int. Trans. Electr. Energy Syst.* **2014**, *25*, 1319–1332. [CrossRef]

8.   Wilson, P.R.; Ross, J.N.; Brown, A.D. Simulation of magnetic component models in electric circuits including dynamic thermal effects. *IEEE Trans. Power Electron.* **2002**, *17*, 55–65. [CrossRef]

9.   Kazimierczuk, M.K. *High-Frequency Magnetic Components*; Wiley: Hoboken, NJ, USA, 2014.

10. Tumański, S. *Handbook of Magnetic Measurements*; CRC Press: Boca Raton, FL, USA, 2011.

11. Detka, K.; Górecki, K. Modelling the power losses in the ferromagnetic materials. *Mater. Sci-Pol.* **2017**, *35*, 398–404. [CrossRef]

12. Barlik, R.; Nowak, M. *Energoelektronika Elementy Podzespoły Układy*; Politechnika Warszawska: Warszawa, Poland, 2014.

13. FERYSTER Sp. z o.o. Sp.k. Toroidal ferrite cores (RTF type). Available online: https://feryster.pl/rdzenie-rtf (accessed on 25 July 2020).

14. Fiorillo, F.; Bertotti, G.; Appino, C.; Pasquale, M. Soft magnetic materials. In *Wiley Encyclopedia of Electrical and Electronics Engineering*; Webster, J.G., Ed.; Wiley: Hoboken, NJ, USA, 2016.

15. Górecki, K.; Detka, K. Application of average electrothermal models in the SPICE-aided analysis of boost converters. *IEEE Trans. Ind. Electron.* **2018**, *66*, 2746–2755. [CrossRef]

16. Górecki, K.; Detka, K. Influence of Power Losses in the Inductor Core on Characteristics of Selected DC–DC Converters. *Energies* **2019**, *12*, 1991. [CrossRef]

17. Detka, K.; Gorecki, K.; Zarebski, J. Modeling single inductor DC–DC converters with thermal phenomena in the inductor taken into account. *IEEE Trans. Power Electron.* **2016**, *32*, 7025–7033. [CrossRef]

18. Gorecki, K.; Gorski, K. Compact thermal model of planar transformers. In Proceedings of the 24th International Conference "Mixed Design of Integrated Circuits and Systems"—Mixdes, 2017, Bydgoszcz, Poland, 22–24 June 2017; pp. 345–350.

19. Benhadda, Y.; Hamid, A.; Lebey, T. Thermal modeling of an integrated circular inductor. *J. Nano-Electron. Phys.* **2017**, *9*, 01004-1–01004-5. [CrossRef]

20. Castellazzi, A.; Gerstenmaier, Y.C.; Kraus, R.; Wachutka, G.K.M. Reliability analysis and modeling of power MOSFETs in the 42-V-PowerNet. *IEEE Trans. Power Electron.* **2006**, *21*, 603–612. [CrossRef]

21. Narendran, N.; Gu, Y. Life of LED-Based White Light Sources. *J. Disp. Technol.* **2005**, *1*, 167–171. [CrossRef]

22. Chang, M.-H.; Das, D.; Varde, P.V.; Pecht, M. Light emitting diodes reliability review. *Microelectron. Reliab.* **2012**, *52*, 762–782. [CrossRef]

23. Górecki, K.; Gorecki, P.; Zarebski, J. Measurements of parameters of the thermal model of the IGBT module. *IEEE Trans. Instrum. Meas.* **2019**, *68*, 4864–4875. [CrossRef]

24. Bagnoli, P.E.; Casarosa, C.; Ciampi, M.; Dallago, E. Thermal resistance analysis by induced transient (TRAIT) method for power electronic devices thermal characterization—Part I: Fundamentals and theory. *IEEE Trans. Power Electron.* **1998**, *13*, 1208–1219. [CrossRef]

25. Sofia, J.W. Analysis of thermal transient data with synthesized dynamic models for semiconductor devices. *IEEE Trans. Compon. Packag. Manuf. Technol. Part A* **1995**, *18*, 39–47. [CrossRef]

26. Masana, F.N. Extraction of structural information from thermal impedance measurements in time domain. In Proceedings of the 18th International Conference Mixed Design of Integrated Circuits and Systems, MIXDES 2011, Gliwice, Poland, 16–18 June 2011; pp. 398–402.

27. D'Alessandro, V.; Rinaldi, N. A critical review of thermal models for electro-thermal simulation. *Solid-State Electron.* **2002**, *46*, 487–496. [CrossRef]

28. Yener, Y.; Kakac, S. *Heat Conduction*; Taylor & Francis: Boca Raton, FL, USA, 2008.

29. Székely, V. A new evaluation method of thermal transient measurement results. *Microelectron. J.* **1997**, *28*, 277–292. [CrossRef]

30. Górecki, K.; Zarębski, J. Nonlinear compact thermal model of power semiconductor devices. *IEEE Trans. Components Packag. Technol.* **2010**, *33*, 643–647. [CrossRef]

31. Janicki, M.; Torzewicz, T.; Samson, A.; Raszkowski, T.; Napieralski, A. Experimental identification of LED compact thermal model element values. *Microelectron. Reliab.* **2018**, *86*, 20–26. [CrossRef]

32. Górecki, K. Modelling mutual thermal interactions between power LEDs in SPICE. *Microelectron. Reliab.* **2015**, *55*, 389–395. [CrossRef]

33. Górecki, K.; Detka, K.; Górski, K. Compact Thermal model of the pulse transformer taking into account nonlinearity of heat transfer. *Energies* **2020**, *13*, 2766. [CrossRef]

34. Janicki, M.; Torzewicz, T.; Ptak, P.; Raszkowski, T.; Samson, A.; Górecki, K. Parametric Compact Thermal Models of Power LEDs. *Energies* **2019**, *12*, 1724. [CrossRef]

35. Gorecki, K.; Gorecki, P. A new form of the non-linear compact thermal model of the IGBT. In Proceedings of the 12th IEEE International Conference on Compatibility, Power Electronics and Power Engineering, Doha, Qatar, 10–12 April 2018. [CrossRef]

36. Górecki, K.; Górski, K.; Zarębski, J. Investigations on the influence of selected factors on thermal parameters of impulse-transformers. *Inf. MIDEM-J. Microelectron. Electron. Compon. Mater.* **2017**, *47*, 3–13.

37. Górecki, K.; Detka, K.; Górski, K. The nonlinear compact thermal model of the pulse transformer. In Proceedings of the 25 International Workshop on Thermal Investigations of ICs and Systems Therminic 2019, Lecco, Italy, 25–27 September 2019. [CrossRef]

38. Ptak, P.; Górecki, K.; Dziurdzia, B. Modelling thermal properties of large LED modules. *Mater. Sci.* **2019**, *37*, 628–638.

39. Christiaens, F.; Vandevelde, B.; Beyne, E.; Mertens, R.; Berghmans, J. A generic methodology for deriving compact dynamic thermal models, applied to the PSGA package. *IEEE Trans. Compon. Packag. Manuf. Technol.* **1998**, *21*, 565–576. [CrossRef]

40. Górecki, K.; Krac, E. Measurements of thermal parameters of solar modules. *J. Physics Conf. Ser.* **2016**, *709*, 012007. [CrossRef]

41. Zarębski, J.; Górecki, K. Parameters estimation of the d.c. electrothermal model of the bipolar transistor. *Int. J. Numer. Model. Electron. Netw. Devices Fields* **2020**, *15*, 181–194. [CrossRef]

42. Górecki, K.; Zarębski, J.; Górecki, P.; Ptak, P. Compact thermal models of semiconductor devices—A review. *Int. J. Electron. Telecommun.* **2019**, *65*, 151–158.

43. Optex Product Information. Available online: http://www.dewetron.cz/optex/katlisty/PT-3S.pdf (accessed on 25 July 2020).

44. F867 Material Characteristics. Available online: http://sklep.remagas.pl/images/W\T1\laściwości_materia\T1\lu_F-867.pdf (accessed on 25 July 2020).

Article

The Influence of an Additional Sensor on the Microprocessor Temperature

Gilbert De Mey [1] and Andrzej Kos [2],*

[1] Department of Electronics and Information Systems, University of Ghent, Technologiepark Zwijnaarde 126, 9052 Zwijnaarde, Belgium; Gilbert.DeMey@UGent.be
[2] Institute of Electronics, AGH University of Science and Technology, Al. Mickiewicza 30, 30-059 Cracow, Poland
* Correspondence: kos@agh.edu.pl

Received: 12 May 2020; Accepted: 17 June 2020; Published: 18 June 2020

Abstract: This paper deals with the problem of inserting a temperature sensor in the neighbourhood of a chip to monitor the junction temperature. If the sensor is not in the middle of the heat source, the recorded temperature can be quite different from the chip temperature we are mainly interested in. For the steady state temperature, it is rather easy to introduce a correction factor. For the transient behaviour of the temperature, there is a tremendous difference between the chip and the sensor temperature, which cannot be neglected if the temperature is used as a parameter to change, for example, the clock frequency in order to improve the throughput.

Keywords: temperature sensors; microprocessor; throughput improvement

---

1. Introduction

Thermal management is a crucial issue in high speed data processing today. The problem has been attacked by many authors, e.g., they propose three techniques to create sensor infrastructures for monitoring the maximum temperature of a multicore system [1]. In [2], the systematic techniques for determining the optimal locations for thermal sensors to provide high-fidelity thermal monitoring of a complex microprocessor system are presented. Another paper presents a compact thermal model that can be integrated with modern Computer Aided Design tools to achieve a temperature-aware design methodology [3].

Data processing with the use of electron devices involves heat losses, which hamper the speed of the processing. Some kinds of cooling systems are used; however, some of them consume additional energy, make noise and enlarge the dimensions of mobile devices. In previously published papers, the authors presented a new idea: one additional temperature sensor, placed on the heat sink. Measuring the temperature difference between the processor and heat sink yields valuable information, which is able to improve the microprocessor's throughput without any changes in its design [4,5]. The theoretical research of these articles was supplemented with experiments using a portable computer: MSI U270 (Micro-Star International Co., Ltd).

It must also be stressed that in this paper, we are dealing with transient or time-dependent thermal problems. Some time ago, there was only interest in steady-state temperatures. Data books only provided thermal resistance. For the power consumption, only a Direct Current (DC) value was given. Recently, more and more attention is paid to transient analyses [6–12]. From these measurements, one can set up equivalent Foster and Cauer Resistor-Capacitor (RC) networks, also known as structure functions [13]. The time constant distribution has also been proved to give a lot of useful information about the thermal path between the chip and the ambient temperature [14]. Just like for linear electric circuit analysis, the time-dependent analysis was mainly done in the AC domain. It turned out that for thermal problems, the Alternating Current (AC) approach was very useful. These studies have been applied to electronic packages [15–18], cooling fins [19,20], underground and overhead high voltage cables [21–24], integrated inductors [25], heat pipes [26] and photovoltaic panels [27].

The title of this paper clearly mentions "additional temperature sensor". Usually, an integrated circuit is fully designed and, at the last minute, the idea is put forward to include some temperature monitoring. In order to avoid a completely new design, the decision is then made to put the temperature sensor "aside" or in a free place, if available. This creates a particular problem that will be discussed in this paper.

2. Experimental Measurements

Measurements have been carried out on a Samsung RC dual core, including AMD E-350 (AMD, USA). This chip contains a dual core GPU Radeon 6320/6310 C (ATI Technologies, Markham, ON, Canada). Figure 1 shows the activity (or CPU usage) and the temperature of the CPU versus time. The temperature of the GPU is also displayed. The GPU is a different integrated circuit. The chip contains a dual core CPU (2x Bobcat) with a L2 Cache (L2 Cache—it is a level of memory, general meaning), a GPU with DirectX v11 GFX (Microsoft), and RAM DDR3 (standard); its temperature will not be used further on in our analysis. Nevertheless, it is clear that both temperatures have a similar, almost identical behaviour. Note that all the temperatures are temperature rises above ambient temperature, i.e., of the metal base plate connected to the cooling fin through a heat pipe.

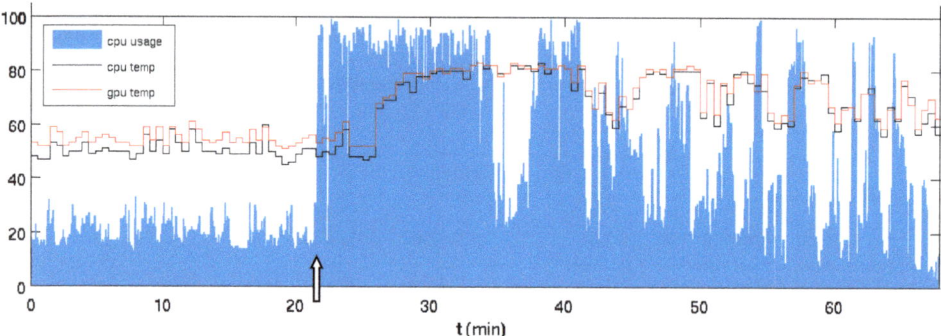

Figure 1. Microprocessor activity and CPU/GPU temperature.

Special software has been installed to measure the activity of the CPU. Activity is measured on a relative scale from 0 to 100. Figure 1 clearly demonstrates the correlation between the activity and the CPU temperature. During the first 21.6 min (=arrow shown in Figure 1), the activity varies around the value of 20. After that period, the activity was increased by a factor 4.5 up to a value around 90. Nevertheless, the temperature rises from 59° to 88° (=59° + 29°) (after extrapolation to steady state). Hence, the temperature is not proportional to the activity. This is due to the technology used for the processor. The power dissipation (and hence the temperature rise) consists of two components: a constant value and a variable component proportional to the clock frequency or, in other words, the activity.

3. Network Model

As shown in Figure 2, the experimental recorded CPU temperature (Figure 1) can be quite well fitted to the following function:

$$T_{fitt}(t) = 58 + 29\left(1 - \frac{210}{210-100}e^{-t/210} + \frac{100}{210-100}e^{-t/100}\right) = 58 + 29\left(1 - 1.909\,e^{-t/210} + 0.909\,e^{-t/100}\right) \quad (1)$$

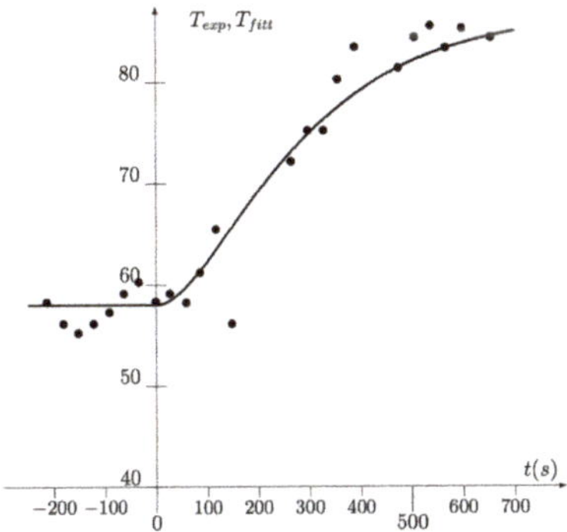

Figure 2. Experimental recorded CPU temperature T_{exp} (•) compared with an analytical function T_{fitt}.

The moment $t = 0$ in Figure 2 is taken as the starting point of the increased activity of the CPU. It corresponds to the moment around 21.6 min (arrow in Figure 1). Note that the Function (1) satisfies the initial conditions $T = 58°$ and $dT/dt = 0$ at the moment $t = 0$. The temperature of $58°$ is the steady state temperature rise during the long reduced activity period ($t < 0$). For modelling, this constant value will no longer be taken into consideration, because we are mainly interested in the transient behaviour.

A closer look at Figure 1 reveals that the start of the increased activity can be approximately seen as a step function of the power consumption. It is then rather straightforward to establish an equivalent network giving rise to a transient temperature like Equation (1), provided a power step is inputted.

A quick look at Figure 1 shows there is a quite long delay between the start of the increased activity and the temperature rise. This proves that the temperature sensor is not in the middle of the heat source but at a certain distance, e.g., on edge of a heat sink. The heat has to propagate a certain time before any temperature rise can be recorded by the sensor. Figure 3 shows the proposed equivalent network. ΔP_0 is the power step due to the increased activity. Note that the node S, where the temperature is evaluated, is not the input node CPU.

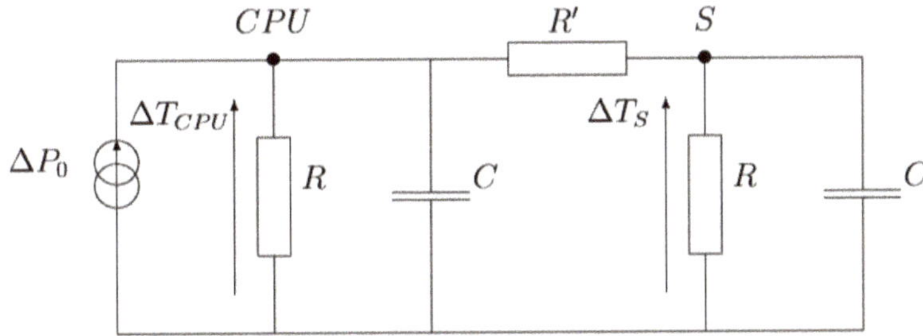

Figure 3. Equivalent thermal network. ΔT_{CPU} and ΔT_S are temperature rises above the reference (=58°).

This is necessary to model the experimentally observed delay. The fact that the two exponential functions in Equation (1) have different signs also proves that we are dealing with transfer impedance. If the temperature would have been calculated at the input node, different signs, like in Equation (1), are physically impossible. Both the CPU and the sensor S are connected to the reference temperature through a thermal resistance R and a thermal capacitance C. We gave them both the same values because there are both within the same package. The coupling resistance R' is responsible for the delay between the CPU and sensor's temperature.

The sensor temperature ΔT_S (Figure 3) is easily found to be given by:

$$\Delta T_S(t) = \Delta P_0 \frac{R^2}{R' + 2R}\left(1 - \frac{R' + 2R}{2R}e^{-t/\tau_1} + \frac{R'}{2R}e^{-t/\tau_2}\right) \tag{2}$$

where ΔP_0 denotes the power step at the input. The method used to get Equation (2) is based on the use of symmetrical components as will be outlined in the appendix further on. In Equation (2), the time constants are given by:

$$\tau_1 = RC, \tau_2 = \frac{RR'}{2R + R'}C \tag{3}$$

The comparison between the experimental fitting, Equation (1), and the network solution, Equation (2), is quite straightforward. One may observe immediately that:

$$\tau_1 = 210 \text{ s}, \quad \tau_2 = 100 \text{ s and } R' = 1.818R \tag{4}$$

Referring to Figure 3, the temperatures ΔT_{CPU} and ΔT_S are in a steady state related by:

$$\Delta T_S = \Delta T_{CPU}\frac{R}{R' + R} = \frac{\Delta T_{CPU}}{2.81} \tag{5}$$

Although the CPU temperature is recorded at S, the displayed value has been multiplied by 2.81 in order to obtain the correct CPU temperature (at least during the steady state). This explains the values of almost 90° in Figure 1 during a high activity period. In order to obtain the correct sensor temperature, the expression (1) should be divided by 2.81.

In steady-state conditions, the total thermal resistance of the CPU to the base plate is the parallel connection of R and $R' + R$:

$$R_{CPU} = \frac{R(R' + R)}{R' + 2R} = \frac{3.81}{4.81}R = 0.792R \tag{6}$$

According to the supplier's information, a maximum temperature rise of 90° is obtained for a power dissipation between 45 and 50 Watts. Taking the average value 47.5 watts, we get:

$$R_{CPU} = \frac{90}{47.5} = 1.8947\frac{K}{W} \tag{7}$$

From Equations (4) and (6), we get then the values of the resistor of the equivalent network.

$$R = 2.392\frac{K}{W} \text{ and } R' = 4.349\frac{K}{W} \tag{8}$$

From the knowledge on the thermal resistance R and the time constant τ_1, one gets the value of the thermal capacitance C:

$$C = \frac{\tau_1}{R} = \frac{210}{2.392} = 87.79\frac{J}{K} \tag{9}$$

A thermal capacitance is, by definition, known as $C = C_v V$, where c_v is the specific heat per unit volume and V, the volume. Most solid materials have a volumetric specific heat of around 2×10^6 J/m^{3}K. Hence, we roughly get:

$$V = \frac{C}{c_v} = \frac{87.79}{210^6} = 43.8 \text{ cm}^3 \tag{10}$$

This seems to be a too high value at first sight. However, one should bear in mind that the chip is connected to the cooling fin through a heat pipe, which is a thermal short circuit. Hence, the volume of 43.8 cm^3 is reasonable if one takes the heat pipe and the cooling fin into account.

4. Discussion

As already mentioned in the foregoing section, the sensor temperature differs from the CPU value that we are mainly interested in. A correction has been made, see Equation (5), so that the correct CPU temperature is obtained in steady-state conditions. However, we want to find out how much difference is obtained in case of a thermal transient problem.

To find an answer to that problem, one can use the equivalent network shown in Figure 3. If a step input power is applied, the sensor temperature ΔT_S is given by Equation (2). The CPU temperature is then found to be:

$$\Delta T_{CPU}(t) = \Delta P_0 \frac{R(R+R')}{R'+2R}\left(1 - \frac{R'+2R}{2(R+R')}e^{-t/\tau_1} - \frac{R'}{2(R+R')}e^{-t/\tau_2}\right) \tag{11}$$

Inserting the numerical values $\tau_1 = 210$ s, $\tau_2 = 100$ s and $R' = 1.818R$ in Equation (11), one gets:

$$\Delta T_{CPU}(t) = 29\left(1 - 0.6774e^{-t/210} - 0.3225e^{-t/100}\right) \tag{12}$$

Notice that in Equation (12), two minus signs appear, which is physically possible because we are dealing with an impedance function, i.e., the temperature is measured between the same nodes where the power is applied.

However, the curve ΔT_S was multiplied by the correction factor 2.81 so that both curves have the same steady state value, so that a better comparison can be made. As can be seen from Figure 4, there is delay of about 100 s between the CPU and the temperature recorded by the sensor. As a consequence, a heat pulse lasting for less than 100 s will not be detected properly by the temperature sensor.

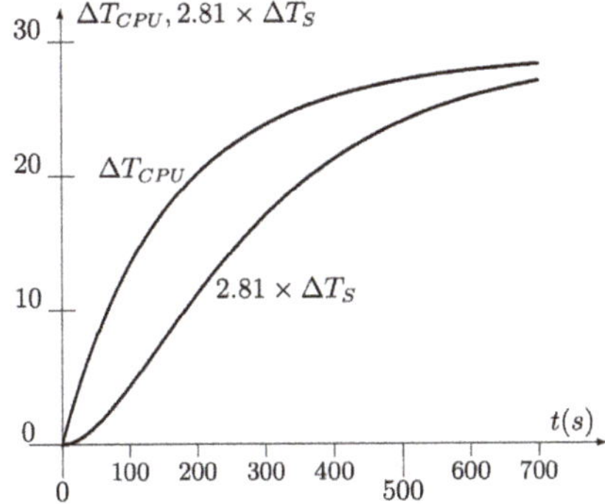

Figure 4. CPU and sensor temperature versus time.

The curve ΔT_{CPU} shows also a sharp rise at $t = 0$, which is typical for the temperature of a heat source. During the transient period, the sensor temperature gives a serious underestimate of the *CPU* temperature and, vice versa, an overestimation of the temperature will occur when the power is suddenly reduced.

It is also remarkable that a lot of information regarding the dynamic thermal properties could be gained from a graph like Figure 1. However, the processor was doing a job that has nothing to do with thermal analysis.

5. Conclusions

The investigation proves that if the temperatures are measured both inside the microprocessor structure and outside of it, e.g., at the cooling fin, the difference between the temperatures gives very useful information. It means that it is possible to increase the power dissipation in the mentioned periods of time without the temperature rising over assumed limit. As a consequence, the microprocessor's throughput increases. Some examples carried out by the authors show that it is possible to improve the throughput even by 7% without any changes in the semiconductor structure, as has been published recently.

However, it has been clearly demonstrated in this paper that it is necessary to put the temperature-sensing device inside the heat source. As soon as the sensor is located at a certain distance, serious errors occur if one wants to measure the transient temperature behaviour. Regarding the steady-state temperature, a simple correction factor can be used, but for a transient problem, the inevitable delay cannot be adjusted by a simple correction factor.

Author Contributions: The conceptualization, formal analysis and methodology presented in this manuscript and their evaluation were carried out by all authors G.D.M. and A.K. The funding acquisition of the financial support for the project leading to this publication was made by A.K. All authors have read and agreed to the published version of the manuscript.

Funding: The research was supported financially from the AGH University of Science and Technology, Krakow, Poland subvention no. 16.16.230.434. The authors also want to express thanks to P. Fluder for his valuable contributions to the experimental measurements.

Conflicts of Interest: The authors declare no conflict of interest.

Appendix A

The easiest way to solve the network of Figure 3 is to use the so-called symmetrical components. One will notice that, apart from the current source, the network shown in Figure 3 is symmetrical.

Hence, the network can be represented as the superposition of two networks (*a* and *b*) shown in Figure A1. On the right-hand side, opposite current sources were introduced so that the total current remained zero. The network of Figure A1a is perfectly symmetrical so that no current can flow through the resistor R'. Hence, the network of Figure A1a is equivalent to the network of Figure A1c. The network of Figure A1b is antisymmetrical, which means that the middle of the resistor R' is always at zero potential. Consequently, this network is equivalent to the network shown in Figure A1d. At last, we just have to solve the networks of Figure A1c,d. Both are simple *RC* networks that can be easily solved by inspection. Immediately, one finds the time constants of Equation (3). Adding the two solutions gives rise to the relation in Equation (2).

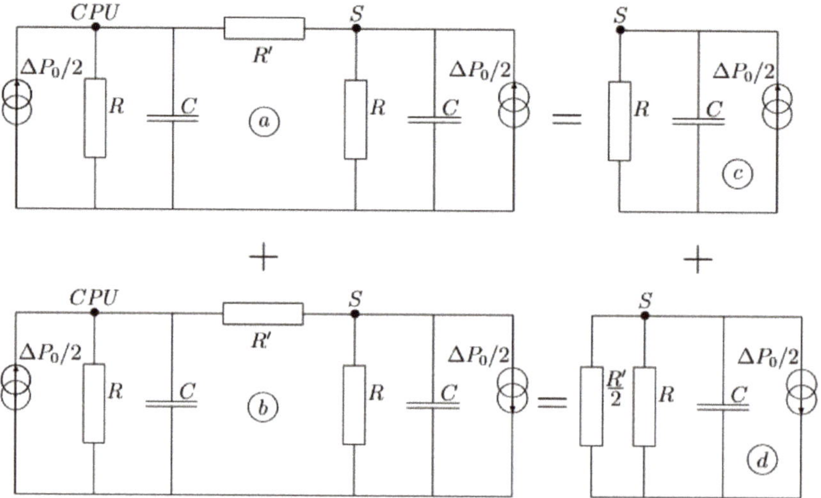

Figure A1. Equivalent thermal network represented as the superposition of a symmetrical and an antisymmetrical network.

References

1. Long, J.; Ogrenci-Memik, S.; Memik, G.; Mukherjee, R. Thermal monitoring mechanisms for chip multiprocessors. *ACM Trans. Arch. Code Optim.* **2008**, *5*, 1–33. [CrossRef]
2. Memik, S.O.; Mukherjee, R.; Ni, M.; Long, J. Optimizing Thermal Sensor Allocation for Microprocessors. *IEEE Trans. Comput. Des. Integr. Circuits Syst.* **2008**, *27*, 516–527. [CrossRef]
3. Huang, W.; Stan, M.R.; Skadron, K.; Sankaranarayanan, K.; Ghosh, S.; Velusam, S. Compact thermal modeling for temperature-Aware design. In Proceedings of the the 41st annual Design Automation Conference, San Diego, CA, USA, 19–23 July 2004; pp. 878–883.
4. Samake, A.; Kocanda, P.; Kos, A. Effective approach of microprocessor throughput enhancement. *Microelectron. Int.* **2019**, *36*, 14–21. [CrossRef]
5. Samake, A.; Kocanda, P.; Kos, A. Improvement of microsystem throughput using new cooling system. *Sci. Lett. Rzeszow Univ. Technol. Electrotech.* **2016**, *XXIV*, 5–15. [CrossRef]
6. Poppe, A.; Szekely, V. Dynamic temperature measurements: Tools providing a look into package and mount structures. *Electron. Cool.* **2002**, *8*, 10–18.
7. Székely, V.; Van Bien, T. Fine structure of heat flow path in semiconductor devices: A measurement and identification method. *Solid-State Electron.* **1988**, *31*, 1363–1368. [CrossRef]
8. Szekely, V. On Representation of infinite-Length distributed RC one-Ports. *IEEE Trans. Circuits Syst.* **1991**, *38*, 711–719. [CrossRef]
9. Simon, D.; Boianceanu, C.; De Mey, G.; Topa, V. Experimental Reliability Improvement of Power Devices Operated under Fast Thermal Cycling. *IEEE Electron Device Lett.* **2015**, *36*. [CrossRef]
10. Simon, D.; Boianceanu, C.; De Mey, G.; Topa, V.; Spitzer, A. Reliability Analysis for Power Devices Which Undergo Fast Thermal Cycling. *IEEE Trans. Device Mater. Reliab.* **2016**, *16*, 336–344. [CrossRef]
11. Janicki, M.; de Mey, G.; Napieralski, A. Transient thermal analysis of multilayered structures using Green's functions. *Microelectron. Reliab.* **2002**, *42*, 1064–1069. [CrossRef]
12. Janicki, M.; De Mey, G.; Napieralski, A. Application of Green's functions for analysis of transient thermal states in electronic circuits. *Microelectron. J.* **2002**, *33*, 733–738. [CrossRef]
13. Janicki, M.; Banaszczyk, J.; De Mey, G.; Kamiński, M.; Vermeersch, B.; Napieralski, A. Dynamic thermal modelling of a power integrated circuit with the application of structure functions. *Microelectron. J.* **2009**, *40*, 1135–1140. [CrossRef]
14. Janicki, M.; Banaszczyk, J.; Vermeersch, B.; De Mey, G.; Napieralski, A. Generation of reduced dynamic thermal models of electronic systems from time constant spectra of transient temperature responses. *Microelectron. Reliab.* **2011**, *51*, 1351–1355. [CrossRef]
15. Kawka, P.; de Mey, G.; Vermeersch, B. Thermal characterisation of electronics packages using the Nyquist plot of the thermal impedance. *IEEE Trans. Compon. Packag. Technol.* **2007**, *30*, 660–665. [CrossRef]
16. De Mey, G.; Pilarski, J.; Wojcik, M.; Lasota, M.; Banaszczyk, J.; Vermeersch, B.; Napieralski, A. Influence of interface materials on the thermal impedance of electronic packages. *Int. Commun. Heat Mass Transf.* **2009**, *36*, 210–212. [CrossRef]
17. Vermeersch, B.; De Mey, G. Thermal impedance plots of micro-scaled devices. *Microelectron. Reliab.* **2006**, *46*, 174–177. [CrossRef]
18. Vermeersch, B.; De Mey, G. Influence of substrate thickness on thermal impedance of microelectronic structures. *Microelectron. Reliab.* **2007**, *47*, 437–443. [CrossRef]
19. De Mey, G.; Vermeersch, B.; Banaszczyk, J.; Swiatczak, T.; Wiecek, B.; Janicki, M.; Napieralski, A. Thermal impedances of thin plates. *Int. J. Heat Mass Transf.* **2007**, *50*, 4457–4460. [CrossRef]
20. Vermeersch, B.; De Mey, G. A Fixed-Angle Dynamic Heat Spreading Model for (An)Isotropic Rear-Cooled Substrates. *J. Heat Transf.* **2008**, *130*, 121301. [CrossRef]
21. Chatziathanasiou, V.; Chatzipanagiotou, P.; Papagiannopoulos, I.; de Mey, G.; Wiecek, B. Dynamic thermal analysis of underground medium power cables using thermal impedance, time constant distribution and structure function. *Appl. Therm. Eng.* **2013**, *60*, 256–260. [CrossRef]
22. Więcek, B.; De Mey, G.; Chatziathanasiou, V.; Papagiannakis, A.; Theodosoglou, I. Harmonic analysis of dynamic thermal problems in high voltage overhead transmission lines and buried cables. *Int. J. Electr. Power Energy Syst.* **2014**, *58*, 199–205. [CrossRef]

23. Chatzipanagiotou, P.; Chatziathanasiou, V.; De Mey, G.; Więcek, B. Influence of soil humidity on the thermal impedance, time constant and structure function of underground cables: A laboratory experiment. *Appl. Therm. Eng.* **2017**, *113*, 1444–1451. [CrossRef]
24. Papagiannopoulos, I.; Chatziathanasiou, V.; Exizidis, L.; Andreou, G.; de Mey, G.; Wiecek, B. Behaviour of the thermal impedance of buried power cables. *Int. J. Electr. Power Energy Syst.* **2013**, *44*, 383–387. [CrossRef]
25. Kałuża, M.; Więcek, B.; De Mey, G.; Hatzopoulos, A.; Chatziathanasiou, V. Thermal impedance measurement of integrated inductors on bulk silicon substrate. *Microelectron. Reliab.* **2017**, *73*, 54–59. [CrossRef]
26. Legierski, J.; Więcek, B.; De Mey, G. Measurements and simulations of transient characteristics of heat pipes. *Microelectron. Reliab.* **2006**, *46*, 109–115. [CrossRef]
27. De Mey, G.; Wyrzutowicz, J.; De Vos, A.; Marańda, W.; Napieralski, A. Influence of lateral heat diffusion on the thermal impedance measurement of photovoltaic panels. *Sol. Energy Mater. Sol. Cells* **2013**, *112*, 1–5. [CrossRef]

Article

Compact Thermal Modeling of Modules Containing Multiple Power LEDs

Marcin Janicki [1,*], Przemysław Ptak [2], Tomasz Torzewicz [1] and Krzysztof Górecki [2]

[1] Department of Microelectronics and Computer Science, Lodz University of Technology, 90-924 Łódź, Poland; torzewicz@dmcs.pl
[2] Department of Marine Electronics, Gdynia Maritime University, 81-255 Gdynia, Poland; p.ptak@we.umg.edu.pl (P.P.); k.gorecki@we.am.gdynia.pl (K.G.)
* Correspondence: janicki@dmcs.pl; Tel.: +48-42-6312654

Received: 20 May 2020; Accepted: 15 June 2020; Published: 17 June 2020

Abstract: Temperature is an essential factor affecting the operation of light-emitting diodes (LEDs), which are often used in circuits containing multiple devices influencing each other. Therefore, the thermal models of such circuits should take into account not only the self-heating effects, but also the mutual thermal influences among devices. This problem is illustrated here based on the example of a module containing six LEDs forming on the substrate a hexagon. This module is supposed to operate without any heat sink in the natural convection cooling conditions, hence it has been proposed to increase the thermal pad area in order to lower the device-operating temperature. In the experimental part of the paper, the recorded diode-heating curves are processed using the network identification by deconvolution method. This allows for the computation of the thermal time constant spectra and the generation of device-compact thermal models. Moreover, the influence of the thermal pad surface area on the device temperature and the thermal coupling between LEDs is investigated.

Keywords: multi-LED lighting modules; device thermal coupling; compact thermal models

1. Introduction

Nowadays, light-emitting diodes (LEDs) have replaced traditional incandescent light bulbs in virtually all everyday applications, becoming the most frequently used light source [1–3]. Taking into account that their operation involves physical phenomena of different natures, such as electrical, optical or thermal ones, the modeling of these devices calls for a truly multi-domain approach [4–6]. Beyond any doubt, among the main factors influencing the performance of LEDs and affecting their parameters is temperature. Therefore, the accurate prediction of LED junction temperature is crucial for the stability of their lighting parameters and the long life of these light sources [7–11]. Thus, the design and the thermal management of luminaires containing LED light sources requires accurate thermal models [12–16]. Moreover, LED light sources are often manufactured also as modules including drivers consisting of power transistors on the same substrate, hence thermal models of these devices should accurately reflect both self- and mutual heating effects [17–24]. Unfortunately, in most cases, the detailed models of LED modules are not readily available or they have unacceptably long simulation times. Thus, these models are usually realized in a reduced SPICE-like compact form, which could be then easily included in some standard electrical or multi-domain simulators [7,25].

In this paper, the authors present a methodology to generate such compact thermal models based on the practical example of a module containing six power LEDs, which are soldered to an metal core printed circuit board (MCPCB) in the shape of a regular hexagon. This module was developed for specific customer needs and it will serve for the lighting of a worker's operating field at an assembly line, where the intensity of light will be automatically controlled depending on the available daylight [26]. In such applications, the use of multi-LED modules is beneficial because it allows a more accurate distribution of light intensity for lower currents flowing through multiple LEDs. Moreover, taking into account that, because of the limited space, this module in its end-use application is to be cooled only by the means of natural convection without any heat sink, two versions of the module, differing in the size of the thermal pads under the LED packages, were considered.

The following section of this paper introduces the prototype modules, measurement equipment and the adopted research methodology. Then, the temperature measurement results are presented in detail. The acquired experimental data allowed the computation of the thermal structure functions and time constant spectra, hence rendering possible the generation of compact thermal models, which were obtained in the form of Cauer RC ladders. These models were used for the simulations of device heating curves taking into account thermal couplings between the devices. The simulation results were validated against measurements. Finally, the Cauer ladders were converted into their Foster counterparts and their element values were analyzed, demonstrating the important influence of an increased thermal pad area on device temperature.

2. Research Methodology

2.1. Prototype Modules and Measurement Euqipment

The test modules analyzed throughout this paper contained six LEDs forming on the substrate, as presented in Figure 1a, a hexagonal circle with the diameter equal to 27 mm. The investigated white diodes of the XREWHT-L1-0000-006F8 type belonged to the XLamp® family manufactured by Cree. Their maximal forward current value is 0.7 A and their viewing angle is 90°. The dimensions of the MCPCB used for these modules were 50 mm × 50 mm × 1.5 mm. These substrates are dedicated for LED applications and according to the datasheet provided by the manufacturer, their thermal conductivity value is equal to 2 W/(m·K). In order to investigate the influence of the thermal pad area on the LED temperature, the modules were fabricated in two different versions. The first one had the standard size of diode thermal pads, i.e., equal to 6.46 mm × 5.60 mm. Their dimensions, as visible in Figure 1b, were delimited by the LED package width and the spacing between the diode electrodes. The other version of the module had the pad width doubled to 12.92 mm. These large pads are visible in Figure 1a as the light rectangles extending outside the package perimeter.

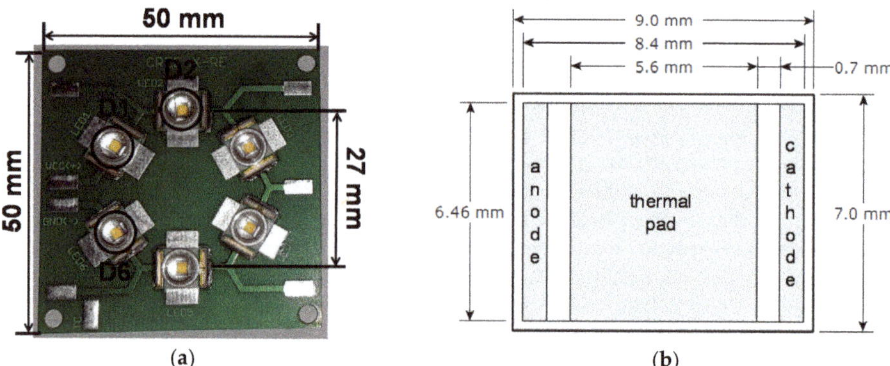

Figure 1. Dimensions of the investigated multi-, light-emitting diodes (LEDs) test module: (**a**) the photograph of the circuit layout; (**b**) the LED electrode layout at the bottom side of the package.

Taking into account the prospective application of the modules, the main goal of the research was to determine the mutual influences between the devices. Looking at the circuit layout, it could be concluded that there existed three possible distances between the devices: 14.1 mm, 24.2 mm or 27.0 mm, therefore it would be enough to include only four LEDs in the thermal coupling analyses. Thus, the diode D2 was initially used as a heat source and the temperature values were measured in the diodes D1–2 and D5–6, but it soon turned out that the responses in the two latter devices were very similar. Hence, the analyses presented in the remainder of this paper will be limited only to the LEDs marked in Figure 1a with black circles. This solution allowed to keep the figures legible without any significant loss of the analysis depth.

The thermal couplings between the devices were investigated based on the results of dynamic temperature measurements, which were taken with the T3Ster® transient thermal tester produced by Mentor®. This piece of equipment renders possible the registration of the system's thermal responses with microsecond time resolution [27]. During the measurements, the diode cathode and anode terminals were soldered to the standard measurement cables provided together with the tester and the MCPCB was placed horizontally in thermally insulated clamps.

2.2. Theoretical Backgound

The entire research methodology employed throughout this paper can be summarized as shown in Figure 2. First, the temperature sensitivity of all LEDs has to be determined and then their thermal responses to the power step thermal excitations have to be recorded. Then, according to the principles of the network identification by deconvolution (NID) method, these responses have to be numerically differentiated so as to obtain the responses to the Dirac delta function and then the deconvolution operation produces the thermal time constant spectra [28], which after segmentation, according to the later described procedure, yield initial compact thermal model element values. Finally, these values are further optimized so as to minimize the simulation errors with respect to the measurements [29].

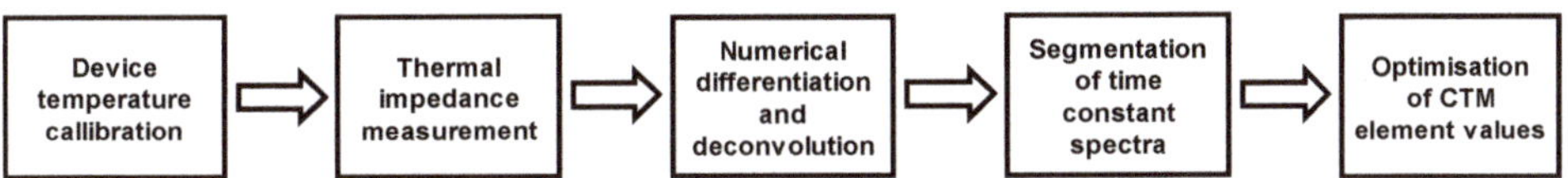

Figure 2. The flowchart illustrating the research methodology employed throughout this paper.

3. Experimental Results

3.1. Device Callibration

Before the actual measurements, each of the investigated devices was calibrated on a cold plate for the forward current of 10 mA. The measured dependencies of diode voltage on temperature are presented in Figure 3a. The black markers and lines correspond to the module having the thermal pads of standard size (STP), whereas the lighter ones to the module with double size thermal pads (DTP). As can be seen in the chart, for each diode this dependence is fairly linear but the measured sensitivity values vary, substantially falling in the range of 1.46÷2.23 mV/K. This result suggests thermal models should be developed independently for each device.

3.2. Transient Thermal Response Measuremnts

This paper will illustrate in detail how the compact thermal model was derived for the diode D2. A similar procedure should be applied for the other LEDs. Initially, during the measurements the constant current of 700 mA was forced through this diode until the thermal steady state was reached, and then, after switching it off, transient temperature responses were measured in the heating device and in the diodes D1 and D6. The measurement current was the same as during the calibration, i.e., 10 mA.

The measurement results are represented in Figure 3b as heating curves, which were obtained by subtracting the recorded diode cooling curves from the respective steady state temperature rise values. The black color in this figure denotes the module with the standard thermal pads. The solid, dashed and double lines are used for the diodes D2, D1 and D6, respectively. The forced LED heating current value was equal to 700 mA, and the measured electrical power was 2.23 W with the standard pad, and 2.18 W with the double one.

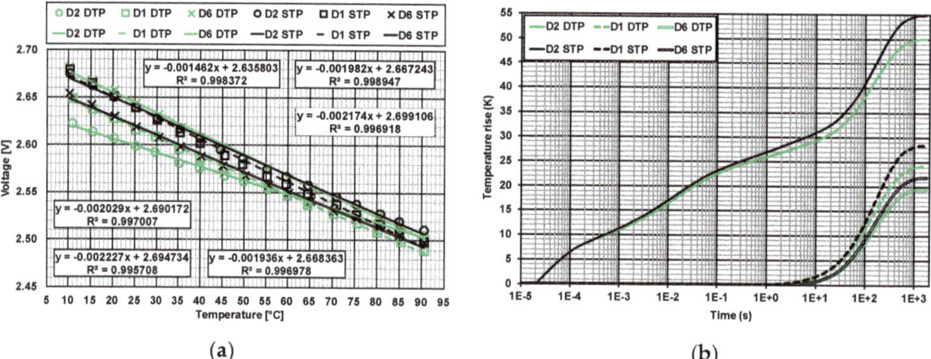

Figure 3. LED measurement results: (**a**) the LED voltage temperature sensitivity; (**b**) the diode heating curves at 700 mA current.

The thermal response in the heating device, as can be seen in the figure, develops almost immediately. The heat diffuses to the other diodes already within a few seconds, reaching 10% of the final temperature rise value in remote devices after more than 20 s. Although the thermal responses in these devices are visibly attenuated, their steady state temperature rise values amount to at least 40% of the maximal temperature rise in the heating diode, hence indicating the existence of the important thermal coupling between the diodes. The measured steady state temperature values in the devices with the larger thermal pads are always at least 10% lower and the influence of the pad in the heating diode becomes visible already after 100 ms. The effect of a larger thermal pad becomes visible already when heat diffuses into the MCPCB in less than a second. This is probably due to the fact that the thermal pad works similarly as the traditional heat spreader, which facilitates the heat removal from the package.

3.3. Computation of Thermal Structure Functions and Time Constant Spectra

According to the JEDEC standards [30,31], all thermal analyses should be always carried out using the real heating power as the input quantity, which can be found by subtracting from the measured electrical power the value of the optical power emitted in the form of light. Thus, it was also necessary to determine the value of the LED optical power, which could be computed using the method described in [32], based on the knowledge of the measured light intensity at a known distance directly over the diode and the spatial light distribution curve provided by the manufacturer in the datasheet. For the LED heating current considered here, the measured optical and real heating power values were equal to 0.4 W and 1.8 W. Unfortunately, without the information concerning the internal diode structure, it was not possible to evaluate how much heat was dissipated in the semiconductor structure itself and how much was dissipated during the wavelength conversion in phosphorus.

Once knowing the LED heating power, the measured curves were processed using the network identification by deconvolution (NID) method offering the entire set of different thermal analysis tools [28]. The thermal cumulative structure functions and the time constant spectra computed using this method are presented in Figure 4. The cumulative thermal structure functions in Figure 4a show the entire heat flow path from the junction (the origin of the co-ordinate system) to the ambient (the steep vertical line at the end). The deflection points in these curves indicate the heat diffusion to another material, and the horizontal plateaus indicate the total accumulated thermal capacitance.

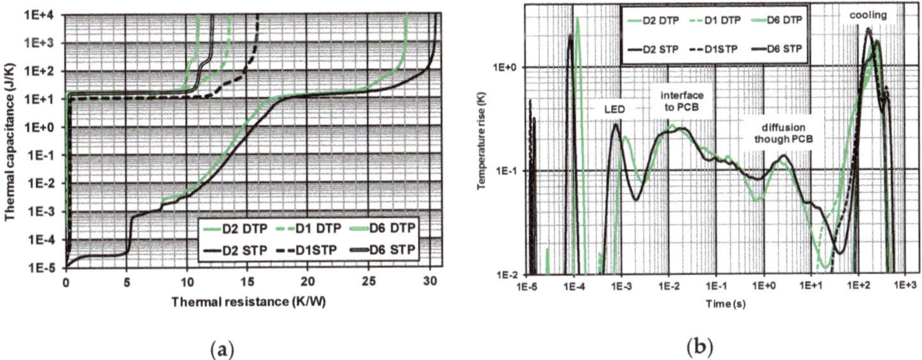

(a)   (b)

Figure 4. Thermal analysis results: (**a**) the cumulative structure functions; (**b**) the time constant spectra.

Normally, the structure functions are computed only for the driving point thermal impedance, i.e., diode D2 in this case, but here we also included the curves computed for the remote diodes, which have only the large plateau corresponding to the PCB capacitance. The curves for the heating diode D2 have three distinct flat sections, which most probably correspond to the LED die, package and board capacitances. The influence of the larger thermal pad only becomes visible when heat diffuses into the substrate, i.e., for the thermal capacitance over 1 mJ/K and for a resistance of around 8 K/W. All the lighter curves at their ends are shifted visibly to the left of their black counterparts, what confirms the earlier observation that the use of a large thermal pad can effectively reduce the total thermal resistance.

The time constant spectra of the thermal responses computed using the NID method are presented in Figure 4b. In order to expose the short time constant components, the spectra presented here were initially integrated over time. As can be seen, the spectra in the figure contain large thermal time constant components related to the heat exchange with ambient, visible large peaks located around 200 s. Short-time constants are present only in the temperature responses of the heating diode. This part of the heat flow path can be analyzed by dividing the spectra into individual sections in the locations of the minima. These sections are additionally indicated in the figure by the text labels. Consequently, the low thermal resistance section from just below a second to around half a minute reflects the conduction of the heat through the MCPCB. Furthermore, the peaks with the maxima in the range of 10÷30 ms correspond to the interface between the package and the board. The remaining part of the spectra below the minima located at a couple of milliseconds describe the heat flow inside the LED package, except for the narrow peaks visible at tens of microseconds, which are just the artefacts after the removal of the electrical transients from thermal responses.

3.4. Foster RC Ladder Compact Thermal Models

For the generation of the circuit compact thermal models (CTMs), the time constant spectra for the heating LED were divided, as previously discussed, into four individual segments corresponding to the diode package, the interface to the MCPCB, the conduction through the board and the heat exchange with the ambient. This procedure yielded CTMs in the form of four stage Foster RC ladder models. In the case of other LEDs, the models had only one RC stage, which will then be used for the computation of mutual thermal couplings.

Consistent with the method described in [29], initially, the thermal resistor values in all the CTM stages were determined by separately accumulating the respective thermal resistances in each section. Then, the time constant values were found by minimizing the simulation errors with respect to the measured temperature values within each time constant range. Finally, the capacitor values were computed by dividing the respective time constant and resistor values. The heating curves obtained with these models for all the considered diodes in the case of the test module with the standard size thermal pads are compared with the measured values in Figure 5. The lighter lines (simulated (SIM)) are used for the simulated values whereas the measured ones (MES) are represented by the black ones.

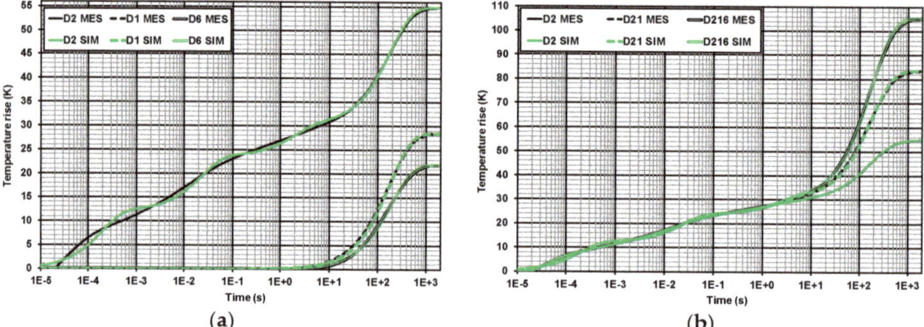

Figure 5. Comparison of measured (MES) and simulated (SIM) heating curves for the LED module with the standard size thermal pads: (**a**) in the different diode locations when only diode D2 is heating; and (**b**) in diode D2 for the different number of heating diodes.

When power was dissipated only in the diode D2, see Figure 5a, the simulation results were very accurate for the time instants over 1 s, where the errors did not exceed 0.7 K. These differences in the time range from 0.1 ms to 1 ms even reached 2.5 K, but the simulation accuracy could be further improved by dividing the first section of the CTM into more segments corresponding to the two distant peaks in the time constant spectra in this region, or by changing the method used to remove the electrical transients from the registered thermal responses. The generated CTMs were also used for simulations when the module was heated by more than one diode, i.e., diodes D2 and D1 (denoted in Figure 5b as D21), and then with the diode D6 (marked as D216) as the third heat source. As can be seen, the thermal coupling between the devices becomes visible only after a few seconds, but the two remote LEDs contribute 50 K to the overall temperature rise of diode D2, which is almost the same value as due to the self-heating.

Generally, the diode heating curves presented in Figure 5 can be generated employing the formula given in Equation (1), where the first component represents the four stage RC model of the *k*-th diode and the second component describes the mutual heating by neighboring devices with the one-stage CTMs. *P* denotes here the respective heating powers of each device:

$$T_k(t) = P_k \sum_{i=1}^{4} R_{ki}\left(1 - \exp(-t/\tau_{ki})\right) + \sum_{j=1}^{5} P_j R_j\left(1 - \exp\left(-t/\tau_j\right)\right) \tag{1}$$

3.5. Cauer Element Values

The previously presented heating curves were obtained with the Foster ladder CTMs because of the straightforward implementation of the mathematical formulas to compute thermal responses. However, as demonstrated in [33], the element values of these models cannot have any physical interpretation because they contain a chain of capacitors linking the diode junction to the ambient, which implies an immediate propagation of thermal responses through the entire module. Therefore, the CTMs were converted to the mathematically equivalent Cauer ladders, which have all their capacitors connected to the thermal ground, so their element values can have some physical interpretation. The conversion was carried out using the algorithm proposed in [34]. The CTM element values for the modules with the standard thermal pad and the doubled one are given in Tables 1 and 2 respectively.

Table 1. Cauer ladder element values for the standard thermal pad.

Device	τ (s)	Rth (K/W)	Cth (J/K)
D2	1.951×10^{-4}	6.837×10^{0}	2.853×10^{-5}
	2.326×10^{-2}	6.403×10^{0}	3.633×10^{-3}
	2.020×10^{0}	3.851×10^{0}	5.245×10^{-1}
	1.759×10^{2}	1.347×10^{1}	1.306×10^{1}
D1	1.805×10^{2}	1.597×10^{1}	1.131×10^{1}
D6	1.960×10^{2}	1.224×10^{1}	1.601×10^{1}

Table 2. Cauer ladder element values for the double thermal pad.

Device	τ (s)	Rth (K/W)	Cth (J/K)
D2	1.819×10^{-4}	6.676×10^{0}	2.725×10^{-5}
	2.090×10^{-2}	6.063×10^{0}	3.446×10^{-3}
	1.181×10^{0}	3.259×10^{0}	3.623×10^{-1}
	1.714×10^{2}	1.218×10^{1}	1.407×10^{1}
D1	1.716×10^{2}	1.366×10^{1}	1.256×10^{1}
D6	1.896×10^{2}	1.105×10^{1}	1.716×10^{1}

The largest time constants of around 3 min contribute over 40% of the thermal resistance in the heating device, and they correspond to the heat exchange with the ambient. These time constants are the only ones present in the responses of the remote diodes, and as mentioned before, the one-stage CTMs generated for these remote diodes can be used for simulations of the thermal couplings between the heating diode and other LEDs, as it was proposed for multi-chip modules in [35]. The time constant around a couple of seconds reflects the heat conduction through the MCPCB and it contributes the least portion to the total thermal resistance. The time constant of just over 20 ms models the interface between the package and the board and the thermal capacitance of this stage equal to around 3.5 mJ/K could be possibly attributed to the diode package. Finally, the shortest time constant of just below 200 µs can be associated with the semiconductor die. Then, the resistance is the junction to the solder point one, given in the datasheets (typically 8 K/W for this device), and the thermal capacitance of the die is equal to almost 30 µJ/K. This value corresponds well to the plateau in the thermal structure functions presented in Figure 4a. Comparing the tables for the standard size thermal pad and the doubled one, it is visible that the major difference arises for the time constants around a few seconds, where both the resistance and capacitance values are noticeably lower for the module with the larger thermal pads, hence confirming that the generated heat is more effectively evacuated from the package to the board.

4. Conclusions

The research results presented in this paper demonstrated the existence of important thermal couplings in the modules containing multiple LEDs and cooled by natural convection without a heat sink. In particular, it was experimentally shown that when cooling is poor the diode self-heating can contribute only a small portion to the total device junction temperature rise over the ambient. Thus, the heating by neighboring devices might become more important than the temperature rise due to the heat generated in a particular device. Therefore, the CTMs used for the simulations of such modules should take into account the mutual thermal interactions between the devices when applied cooling is poor. Obviously, this problem will not be so severe, when a proper heat sink is attached, but sometimes this might not be possible because of various design constraints. Another important contribution of this research was to demonstrate that the thermal pad area noticeably influences the junction-to-board thermal resistance.

From the theoretical point of view, this paper illustrated the methodology to derive compact thermal models when power is dissipated in multiple devices. This methodology turned out to be effective and the thermal simulations carried out with the resulting dynamic compact models were accurate. However, it should be underlined that in the considered case, the geometry of the circuit layout was symmetrical and the circuit contained devices of the same type. Thus, in the general case, the reciprocity of mutual thermal influences would not hold. The main limitation of the proposed methodology was that the compact model element values depend on the device operating conditions and that they also vary among individual devices. Therefore, the compact models have to be derived separately for each LED. These differences result not only from different LED temperature sensitivities, but also from different LED soldering thermal resistances and other factors. Another interesting observation from the system designer point of view is that increasing the area of LED thermal pads can effectively reduce the device temperature rise already after a second. In the steady state, it was possible to lower the temperature even by 10% when only one device was active. This effect could be increased when more devices are dissipating power. In further investigations, a more in-depth study of the thermal pad area should be carried out.

Author Contributions: The measurements presented in this manuscript and their evaluation was carried out by P.P. and T.T. The research on the generation of compact models and thermal simulations was done by M.J. Finally, M.J. and K.G. supervised the research and prepared the manuscript. All authors have read and agreed to the published version of the manuscript.

Funding: This research was supported financially from the Ministry of Science and Higher Education program "Regional Excellence Initiative" 2019-2022 project No. 006/RID/2018/19, the sum of financing 11,870,000 PLN.

Conflicts of Interest: The authors declare no conflict of interest.

References

1. Weir, B. Driving the 21st century's lights. *IEEE Spectr.* **2012**, *49*, 42–47. [CrossRef]
2. Schubert, E.F. *Light Emitting Diodes*, 3rd ed.; Rensselaer Polytechnic Institute: Troy, NY, USA, 2018.
3. Lasance, C.J.M.; Poppe, A. (Eds.) *Thermal Management for LED Applications*; Springer: Dordrecht, The Netherlands, 2014.
4. Poppe, A.; Farkas, G.; Gaál, L.; Hantos, G.; Hegedüs, J.; Rencz, M. Multi-domain modelling of LEDs for supporting virtual prototyping of luminaires. *Energies* **2019**, *12*, 1909. [CrossRef]
5. Martin, G.; Marty, C.; Bornoff, R.; Poppe, A.; Onushkin, G.; Rencz, M.; Yu, J. Luminaire digital design flow with multi-domain digital twins of LEDs. *Energies* **2019**, *12*, 2389. [CrossRef]
6. Poppe, A. Simulation of LED based luminaires by using multi-domain compact models of LEDs and compact thermal models of their thermal environment. *Microelectron. Reliab.* **2017**, *72*, 65–74. [CrossRef]
7. Biber, C. LED Light Emission as A Function of Thermal Conditions. In Proceedings of the 24th IEEE Semiconductor Thermal Measurement and Management Symposium, San Jose, CA, USA, 16–20 March 2008; pp. 180–184. [CrossRef]
8. Narendran, N.; Gu, Y. Life of LED-based white light sources. *J. Disp. Technol.* **2005**, *1*, 167–171. [CrossRef]
9. Chang, M.H.; Das, D.; Varde, P.V.; Pecht, M. Light emitting diodes reliability review. *Microelectron. Reliab.* **2012**, *52*, 762–782. [CrossRef]
10. Song, B.M.; Han, B.; Lee, J.H. Optimum design domain of LED-based solid state lighting considering cost, energy consumption and reliability. *Microelectron. Reliab.* **2013**, *53*, 435–442. [CrossRef]
11. Efremov, A.A.; Bochkareva, N.I.; Gorbunov, R.I.; Lavrinovich, D.A.; Rebane, Y.T.; Tarkhin, D.V.; Shreter, Y.G. Effect of the Joule heating on the quantum efficiency and choice of thermal conditions for high power blue InGaN/GaN LEDs. *Semiconductors* **2006**, *40*, 605–610. [CrossRef]
12. Huang, S.; Wu, H.; Fan, B.; Zhang, B.; Wang, G. A chip-level electrothermal-coupled design model for high-power light-emitting diodes. *J. Appl. Phys.* **2010**, *107*. [CrossRef]
13. Treurniet, T.; Lammens, V. Thermal management in color variable multi-chip LED modules. In Proceedings of the 23rd IEEE Semiconductor Thermal Measurement and Management Symposium SEMI-THERM, Dallas, TX, USA, 14–16 March 2006. [CrossRef]
14. Górecki, K.; Ptak, P. Modeling LED lamps in SPICE with thermal phenomena taken into account. *Microelectron. Reliab.* **2017**, *79*, 440–447. [CrossRef]

15. Alexeev, A.; Onushkin, G.; Linnartz, J.P.; Martin, G. Multiple heat source thermal modeling and transient analysis of LEDs. *Energies* **2019**, *12*, 1860. [CrossRef]
16. Kumar, S.K.; Wani, S.C.; Lee, R. LED thermal management of an automotive electronic control module with display. In Proceedings of the IEEE 14th Electronics Packaging Technology Conference EPTC, Singapore, 5–7 December 2012. [CrossRef]
17. Lan, K.; Moo, W.H. Thermal resistance measurement of LED package with multichips. *IEEE Trans. Compon. Packaging Technol.* **2007**, *30*, 632–636. [CrossRef]
18. Ptak, P.; Górecki, K.; Dziurdzia, B. Modelling thermal properties of large LED module. *Mater. Sci. Pol.* **2019**, *37*, 628–638. [CrossRef]
19. Wang, C.P. Effects of diode voltage and thermal resistance on the performance of multichip LED modules. *IEEE Trans. Electron Devices* **2015**, *63*, 390–393. [CrossRef]
20. Wang, C.P.; Kang, S.W.; Lin, K.M.; Chen, T.T.; Fu, H.K.; Chou, P.T. Analysis of thermal resistance characteristics of power LED module. *IEEE Trans. Electron Devices* **2013**, *61*, 105–109. [CrossRef]
21. Kim, H.; Kim, K.J.; Lee, Y. Thermal performance of smart heat sinks for cooling high power LED modules. In Proceedings of the IEEE 13th Intersociety Conference on Thermal and Thermomechanical Phenomena in Electronic Systems ITHERM, San Diego, CA, USA, 30 May–1 June 2012. [CrossRef]
22. Bahman, A.; Ma, K.; Blaabjerg, F. A lumped thermal model including thermal coupling and thermal boundary conditions for high power IGBT modules. *IEEE Trans. Power Electron.* **2017**, *33*, 2518–2530. [CrossRef]
23. Górecki, K.; Górecki, P.; Zarębski, J. Measurements of parameters of the thermal model of the IGBT module. *IEEE Trans. Instrum. Meas.* **2019**, *68*, 4864–4875. [CrossRef]
24. Deng, Z.; Zhao, Z.; Zhang, P.; Li, J.; Huang, Y. Study of the methods to measure the junction-to-case thermal resistance of IGBT modules and press pack IGBTs. *Microelectron. Reliab.* **2017**, *79*, 248–256. [CrossRef]
25. Gorecki, K. Modelling mutual thermal interactions between power LEDs in SPICE. *Microelectron. Reliab.* **2015**, *55*, 389–395. [CrossRef]
26. Ptak, P.; Górecki, K.; Wnuczko, S. Embedded system to control lighting of the office workplace. *Przegląd Elektrotechniczny* **2018**, *94*, 76–79. [CrossRef]
27. Mentor®. Available online: www.mentor.com/products/mechanical/micred/t3ster/ (accessed on 15 May 2020).
28. Szekely, V. A new evaluation method of thermal transient measurement results. *Microelectron. J.* **1997**, *28*, 277–292. [CrossRef]
29. Janicki, M.; Torzewicz, T.; Samson, A.; Raszkowski, T.; Napieralski, A. Experimental identification of LED compact thermal model element values. *Microelectron. Reliab.* **2018**, *86*, 20–26. [CrossRef]
30. JEDEC. *Implementation of the Electrical Test Method for the Measurement of Real Thermal Resistance and Impedance of Light-Emitting Diodes with Exposed Cooling Surface*; Standard JESD51-51; JEDEC: Arlington, VA, USA, 2012.
31. JEDEC. *Guidelines for Combining CIE 127-2007 Total Flux Measurement with Thermal Measurement of LEDs with Exposed Cooling Surface*; Standard JESD51-52; JEDEC: Arlington, VA, USA, 2012.
32. Janicki, M.; Torzewicz, T.; Ptak, P.; Raszkowski, T.; Samson, A.; Górecki, K. Parametric compact thermal models of power LEDs. *Energies* **2019**, *12*, 1724. [CrossRef]
33. Janicki, M.; Napieralski, A. Considerations on Electronic System Compact Thermal Models in The Form of RC Ladders. In Proceedings of the 15th International Conference the Experience of Designing and Application of CAD Systems CADSM, Polyana, Ukraine, 26 February–2 March 2019.
34. Gerstenmaier, Y.C.; Kiffe, W.; Wachutka, G. Combination of Thermal Subsystem Modeled by Rapid Circuit Transformation. In Proceedings of the 13th International Workshop on Thermal Investigation of ICs and Systems, Budapest, Hungary, 17–19 September 2007; pp. 115–120. [CrossRef]
35. Poppe, A.; Zhang, Y.; Wilson, J.; Farkas, G.; Szabo, P.; Parry, J.; Rencz, M.; Szekely, V. Thermal measurement and modeling of multi-die packages. *IEEE Trans. Compon. Packag. Technol.* **2009**, *32*, 484–491. [CrossRef]

Article

Electrothermal Averaged Model of a Diode-Transistor Switch Including IGBT and a Rapid Switching Diode

Paweł Górecki and Krzysztof Górecki *

Department of Marine Electronics, Gdynia Maritime University, Morska 83, 81-225 Gdynia, Poland; p.gorecki@we.umg.edu.pl
* Correspondence: k.gorecki@we.umg.edu.pl

Received: 18 May 2020; Accepted: 10 June 2020; Published: 12 June 2020

Abstract: This study proposes an electrothermal averaged model of the diode–transistor switch including insulated gate bipolar transistor (IGBT) and a rapid switching diode. The presented model has the form of subcircuits dedicated for simulation program with integrated circuit emphasis (SPICE) and it makes it possible to compute characteristics of DC–DC converters at the steady state considering self-heating phenomena, both in the diode and in IGBT. This kind of model allows computations of voltages, currents and internal temperatures of all used semiconductor devices at the steady state. The formulas used in this model are adequate for both: continuous conducting mode (CCM) and discontinuous conducting mode (DCM). Correctness of the proposed model is verified experimentally for a boost converter including IGBT. Good accuracy in modeling these converter characteristics is obtained.

Keywords: IGBT; DC–DC converter; electrothermal model; averaged model; thermal phenomena; self-heating; diode–transistor switch; power electronics

---

1. Introduction

Nowadays, power electronic circuits often include DC–DC converters [1–5]. Designing these converters requires reliable methods of computer simulations [1,5,6]. They enable selection of optimal components and operating conditions of the designed converter. For more than 50 years, many scientists have been working on methods and models that allow fast computations of waveforms of currents and voltages as well as AC and DC characteristics of power converters [5–8]. To provide simulations of electronic circuits using the formulated models it is necessary to use software dedicated to electronic circuits simulation. SPICE (simulation program with integrated circuit emphasis) is commonly used for this purpose [5,6,8–10]. Its main advantage is easy implementation of any compact model of electronic components and devices.

Due to the switching mode of operation of DC–DC converters, their characteristics can be obtained with the sequential analysis of transient processes only. Such analysis requires, as a rule, a long time—and may cause convergence problems of computations [11]. In order to shorten the time of this analysis, the averaged models of DC–DC converters were proposed [5,8,12–15]. These models make it possible to compute characteristics of circuits at the steady state using a DC sweep (DC analysis) [5,8,16]. This is one of the types of the standard analyses realized in SPICE. Typically, a DC analysis needs a short computation time. In averaged models, the average values of voltages in nodes and currents flowing through branches of considered circuits are used instead of the instantaneous values of these quantities.

It is assumed that averaged DC–DC converters models should be formulated in accordance with two steps of their activation in "on" and "off" phases of a switch. In the first step, the transistor is turned on, and simultaneously the diode is turned off. In the second step, the transistor is turned off and simultaneously the diode is turned on. The considered states of operation are described using two subcircuits which depend on each other. For both the mentioned steps of operation, equations describing currents flowing through capacitors and voltages on inductors are formulated. Next, these equations are averaged across period T and compared to zero. Finally, equations describing dependences between average values of currents and voltage at the steady state in the considered DC–DC converter are obtained [5,8].

Averaged models of DC–DC converters have been described in literature for many years [5,8,12–21]. They can be presented in different network forms and they can be used for: transient, AC and DC analysis [5,8]. They also enable obtaining small-signal frequency characteristics, steady-state voltage–current characteristics and waveforms of currents and voltages in considered converters.

As it can be seen, e.g., in [5,8,16], in all single-inductor DC–DC converters a diode–transistor switch can be distinguished. In such a switch a diode and one kind of transistors (BJT, JFET, MOSFET or IGBT) are included. The structure of such a switch, including insulated gate bipolar transistor (IGBT) and a diode, is shown in Figure 1.

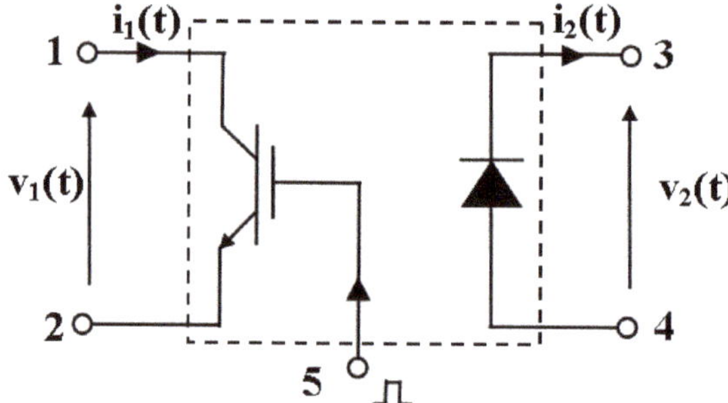

Figure 1. Diagram of a diode–transistor switch with IGBT and the diode.

The considered switch contains 5 terminals: 3 terminals of IGBT and 2 terminals of the diode. During operation of such a switch in any DC–DC converter, a control signal in the form of a rectangular pulses train is connected to terminal number 5. The transistor and the diode are connected to other components of the considered DC–DC converter as shown in its diagram. Voltage between pairs of terminals 1, 2 ($v_1(t)$ voltage) and 3, 4 ($v_2(t)$ voltage) have the shape of rectangular pulses trains. At the steady state waveforms of voltages and currents are periodical.

In many papers, e.g., [5,8,16,17,20,21] averaged models of a diode–transistor switch of different accuracy are described. In some of these models the diode and the transistor are described as ideal switches characterized by zero on-state resistance and infinite off-state resistance [5,8]. Another group of these models uses piecewise linear characteristics of the mentioned semiconductor devices [5,8,16,17,21]. Some of these models [21] take into account operation only in CCM (Continuous Current Mode), whereas models described in [5,8,16] make it possible to compute characteristics of these converters operating both in CCM and DCM (Discontinuous Current Mode).

During operation of DC–DC converters self-heating phenomena occur in semiconductor devices contained in these converters [11,16,17,22–24]. The result of these phenomena is an increase in internal temperature of these devices and changes in the course of characteristics of these converters. As it is indicated, among others in [11,17], thermal phenomena occurring in components of low-voltage converters significantly influence characteristics of these converters. Thermal phenomena also strongly affect reliability of such converters, as presented in [11,16,17]. A computer analysis can be useful in estimating thermal phenomena. In order to apply such a method, an electro-thermal model of electronic components comprising a DC–DC converter is necessary.

In literature, averaged models of a diode–transistor switch containing ideal switches [5,8,18,20] or MOSFETs [16,17,25] are described. The model of a diode–transistor switch containing IGBT and a diode presented in the study [21] has a significantly simplified form. In the considered model, characteristics of the diode and IGBT are modeled using piecewise linear functions consisting of only two pieces. Such a manner of modeling DC characteristics of IGBT and a diode can cause a significant error of computations, especially in the range of low currents. Moreover, this model was verified only in the CCM. As it is pointed out in many papers [5,8,16], omitting DCM can also constitute a cause of a significant error in computations. Finally, in the model presented in [21] thermal phenomena are not taken into account. The electrothermal averaged model of a diode–transistor switch including the power MOS transistor is described in [16,17]. Such models enable computations of voltages and currents in the considered converter and internal temperatures of semiconductor devices contained in the diode–transistor switch. Due to different shapes of output characteristics of IGBT and MOSFET operating in the on–state, the cited models are not adequate to analyze characteristics of DC–DC converters including IGBT and a diode.

In this study, the electrothermal averaged model of a diode–transistor switch including the IGBT is proposed. It is dedicated to electrothermal analysis of DC–DC converters containing such a switch and it makes it possible to compute values of current, voltage and internal temperatures of semiconductor devices contained in the mentioned switch. In electrothermal analyses, both electrical phenomena occurring in the analyzed DC–DC converter as well as self-heating phenomena occurring in both semiconductor devices are taken into account. In Section 2, the form of the elaborated model is presented. In Section 3, a manner of estimating model parameters is described and some results of modeling DC characteristics of components of the diode–transistor switch are shown. In Section 4, the results of usefulness of the elaborated model for determining characteristics of the boost converter are presented and discussed.

2. Proposed Model

The formulated electrothermal model of a diode–transistor switch is based on the concept described in the study [16,17] for such a switch including the power MOS transistor. The network representation of the proposed electrothermal model of a diode–transistor switch (DTS) with IGBT and a diode is shown in Figure 2.

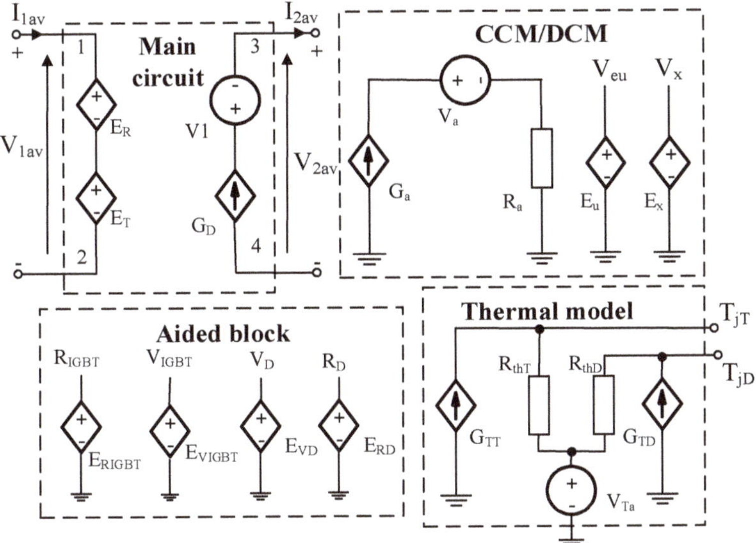

Figure 2. Network representation of the electrothermal averaged model of a diode–transistor switch (DTS) including IGBT and a diode.

Terminals of the presented model denoted as 1, 2, 3 and 4 correspond to terminals of the real diode–transistor switch in the diagram shown in Figure 1. This model can be connected to other parts of the analyzed DC–DC converters through the mentioned terminals in accordance with the diagram of these converters. The control signal is not shown in the model, but frequency f and duty cycle d of this signal are parameters of the presented model that are used in descriptions of some controlled voltage and current sources existing in this model. The voltages on terminals T_{jT} and T_{jD} correspond to internal temperatures of IGBT and the diode, respectively.

The presented model consists of 4 blocks. These are: main circuit, aided block, thermal model and CCM/DCM. Blocks and all components contained in them are described below.

The main circuit includes two controlled voltage sources E_R and E_T modeling average voltage between output terminals of IGBT and controlled current source G_D modeling the average value of diode current. Voltage source V1 of zero value is used to monitor the average value of diode current. Values of currents I_1av and I_2av, as well as voltages V_1av and V_2av in the main circuit, are average values of diode and transistor voltages and currents. The connected in series controlled voltage sources E_R and E_T model a voltage drop between transistor output terminals. Controlled current source G_D models diode current. The output values of these sources are described with elaborated formulas using the concept presented in the papers [16,17]. According to this concept, in each period of the control signal, current flows through the transistor for a time equal to d·T, and through the diode—for a time equal to (1-d)·T, wherein T is period of the control signal. Diode current–voltage characteristic and transistor output characteristic are described by piecewise linear functions. Parameters describing such piecewise linear functions depend on internal temperatures of the diode and IGBT. Values of these parameters are computed in the aided block, whereas internal temperatures of semiconductor devices are computed in the thermal model. The equivalent duty cycle V_{eu}, adequate for operation of the DC–DC converter containing the modeled diode–transistor switch, is computed in CCM/DCM block shown in Figure 2. The value of V_{eu} depends on duty cycle and frequency of the control signal, inductance of the inductor contained in the analyzed DC–DC converter and output current of this converter.

Controlled voltage and current sources existing in the main circuit are described as follows

$$E_T = \frac{1 - V_{eu}}{V_{eu}} \cdot (V_{2av} + V_D) \tag{1}$$

$$E_R = \frac{V_{IGBT}}{V_{eu}} + \frac{I_{1av} \cdot R_{IGBT}}{V_{eu}} + \frac{I_{1av} \cdot R_D}{V_{eu}^2} \cdot (1 - V_{eu}) \tag{2}$$

$$G_D = \frac{1 - V_{eu}}{V_{eu}} \cdot I_{1av} \tag{3}$$

where current I_{1av} and voltage V_{2av} are denoted in Figure 2, V_{eu} is voltage on controlled voltage source E_u, V_D voltage and R_D resistance are described by a piecewise linear characteristic of forward biased diode, whereas V_{IGBT} voltage and resistance R_{IGBT} are described by a piecewise linear output characteristic of switched-on IGBT. Parameters V_{IGBT}, V_D, R_D and R_{IGBT} depend on the device internal temperature (T_{jT} for IGBT and T_{jD} for the diode) and on current flowing through these devices. Values of the mentioned parameters correspond to voltages on controlled voltage sources E_{VIGBT}, E_{VD}, E_{RD} and E_{RIGBT}, respectively. These sources are included in the aided block. Descriptions of the mentioned parameters are as follows:

$$V_D = \begin{cases} V_{D1} \cdot \left(1 + \left(T_{jD} - T_0\right) \cdot \alpha_{VD1}\right) \; if \; I_{2av} < a_1 \\ V_{D2} \cdot \left(1 + \left(T_{jD} - T_0\right) \cdot \alpha_{VD2}\right) \; if \; a_1 \leq I_{2av} < a_2 \\ V_{D3} \cdot \left(1 + \left(T_{jD} - T_0\right) \cdot \alpha_{VD3}\right) \; if \; I_{2av} \geq a_2 \end{cases} \tag{4}$$

$$R_D = \begin{cases} R_{D1} \cdot \left(1 + \left(T_{jD} - T_0\right) \cdot \alpha_{RD1}\right) \; if \; I_{2av} < a_1 \\ R_{D2} \cdot \left(1 + \left(T_{jD} - T_0\right) \cdot \alpha_{RD2}\right) \; if \; a_1 \leq I_{2av} < a_2 \\ R_{D3} \cdot \left(1 + \left(T_{jD} - T_0\right) \cdot \alpha_{RD3}\right) \; if \; I_{2av} \geq a_2 \end{cases} \tag{5}$$

$$V_{IGBT} = \begin{cases} V_{IGBT1} \cdot \left(1 + \left(T_{jT} - T_0\right) \cdot \alpha_{VIGBT1}\right) \; if \; I_{1av} < b_1 \\ V_{IGBT2} \cdot \left(1 + \left(T_{jT} - T_0\right) \cdot \alpha_{VIGBT2}\right) \; if \; b_1 \leq I_{1av} < b_2 \\ V_{IGBT3} \cdot \left(1 + \left(T_{jT} - T_0\right) \cdot \alpha_{VIGBT3}\right) \; if \; I_{1av} \geq b_2 \end{cases} \tag{6}$$

$$R_{IGBT} = \begin{cases} R_{IGBT1} \cdot \left(1 + \left(T_{jT} - T_0\right) \cdot \alpha_{RIGBT1}\right) \; if \; I_{1av} < b_1 \\ R_{IGBT2} \cdot \left(1 + \left(T_{jT} - T_0\right) \cdot \alpha_{RIGBT2}\right) \; if \; b_1 \leq I_{1av} < b_2 \\ R_{IGBT3} \cdot \left(1 + \left(T_{jT} - T_0\right) \cdot \alpha_{RIGBT3}\right) \; if \; I_{1av} \geq b_2 \end{cases} \tag{7}$$

In Equations (5) and (6) symbols V_{D1}, V_{D2}, V_{D3}, R_{D1}, R_{D2}, R_{D3}, α_{VD1}, α_{VD2}, α_{VD3}, α_{RD1}, α_{RD2}, α_{RD3}, a_1 and a_2 denote parameters of a piecewise linear model of diode DC characteristic. In turn, in Equations (6) and (7) symbols V_{IGBT1}, V_{IGBT2}, V_{IGBT3}, R_{IGBT1}, R_{IGBT2}, R_{IGBT3}, α_{VIGBT1}, α_{VIGBT2}, α_{VIGBT3}, α_{RIGBT1}, α_{RIGBT2}, α_{RIGBT3}, b_1 and b_2 denote parameters of a piecewise linear model of IGBT output characteristics. Values of temperatures T_{jD} and T_{jT} are computed in the thermal model, whereas T_0 represents reference temperature.

In the thermal model values of internal temperature of IGBT (T_{jT}) and the diode (T_{jD}) are computed with self-heating phenomena taken into account. The classical electrical analog of a DC compact thermal model described, e.g., in [11,16,17,22,26,27] is used. In this analog temperature corresponds to voltage in selected nodes of this analog, whereas dissipated power is represented by current sources. The ability to remove heat generated in the diode and in IGBT is characterized by thermal resistance. In the proposed model voltage source V_{Ta} represents ambient temperature, resistors R_{thT} and R_{thD} denote thermal resistance of IGBT and the diode, respectively. Average values of power dissipated in the considered semiconductor devices are represented by controlled current sources G_{TT} and G_{TD}. Currents flowing through these sources are described by the following formulas

$$G_{TD} = \left(V_D + \frac{R_D \cdot I_{2av}}{1 - V_{eu}}\right) \cdot \frac{I_{2av}}{1 - V_{eu}} \tag{8}$$

$$G_{TT} = \left(V_{IGBT} + \frac{R_{IGBT} \cdot I_{1av}}{V_{eu}}\right) \cdot \frac{I_{1av}}{V_{eu}} \tag{9}$$

The proposed model can be used for computations of characteristics of DC–DC converters operating in CCM or DCM. In both mentioned modes of operation in each period of a control signal the transistor is turned on in time equal to the product of duty cycle d and period T. In turn, the diode is turned on in time equal to (1-d)·T in CCM and this time is shorter in DCM. In order to take into account influence of duty cycle d, frequency f of the control signal and inductance L of the inductor included in the tested DC–DC converter on voltage V_{eu}, CCM/DCM block is included in the model. This block includes controlled current source G_a, voltage source V_a of zero value, resistor R_a and controlled voltage sources E_u and E_x.

Voltage V_{eu} on voltage source E_u is described by the formula

$$V_{eu} = \begin{cases} 0 \; if \; V_x < 0 \\ V_x \; if \; 0 < V_x \; < 1 \\ 1 \; if \; V_x > 1 \end{cases} \tag{10}$$

where voltage V_x on voltage source E_x is described as follows

$$V_x = LIMIT\left(MAX\left(d, \frac{d^2}{d^2 + 2 \cdot L \cdot f \cdot \frac{I_{Va}}{V_{2av} + V_D}}\right), 0, 1\right) \tag{11}$$

In Equation (11) LIMIT (·) and MAX (·) are SPICE standard functions described, e.g., in the book [28] and I_{Va} denotes current flowing through controlled current source G_a described with the formula of the form:

$$G_a = MAX(I_{1av}, 0) \tag{12}$$

Resistor R_a must be included in this block due to formal rules of SPICE. Voltage source V_a is used to monitor the value of current I_{Va}.

3. Estimation of Model Parameters

Practical applications of the proposed model require estimation of parameter values existing in this model. For example, an estimate is performed for the IGBT of the type IGP06N60T [29] by Infineon Technologies and for the diode of the type IDP08E65 [30] by Infineon Technologies. Input data are characteristics of such devices measured by the authors with the use of the impulse method [31] and a source-meter Keithley 2612a [32] at different fixed values of ambient temperature. When these measurements are performed, the tested semiconductor devices are situated in a thermal chamber, in which the adjustable value of temperature can be obtained.

Values of the model parameters obtained by matching piecewise linear models of characteristics of the considered devices are given in Table 1.

Table 1. Values of parameters of the piecewise linear model of the used insulated gate bipolar transistor (IGBT) and diode.

Parameter Name	V_{D1} (V)	V_{D2} (V)	V_{D3} (V)	α_{VD1} (K^{-1})	α_{VD2} (K^{-1})	α_{VD3} (K^{-1})	a_1 (A)
Parameter Value	0.63	0.74	0.847	-8.41×10^{-3}	-3.92×10^{-3}	-2.95×10^{-3}	0.25
Parameter Name	R_{D1} (Ω)	R_{D2} (Ω)	R_{D3} (Ω)	α_{RD1} (K^{-1})	α_{RD2} (K^{-1})	α_{RD3} (K^{-1})	a_2 (A)
Parameter Value	0.75	0.191	0.1	4.67×10^{-3}	2.62×10^{-3}	1.74×10^{-3}	1.3
Parameter Name	V_{IGBT1} (V)	V_{IGBT2} (V)	V_{IGBT3} (V)	α_{VIGBT1} (K^{-1})	α_{VIGBT2} (K^{-1})	α_{VIGBT3} (K^{-1})	b_1 (A)
Parameter Value	0.611	0.736	0.811	-3.04×10^{-3}	-1.63×10^{-3}	-1.2×10^{-3}	0.52
Parameter Name	R_{IGBT1} (Ω)	R_{IGBT2} (Ω)	R_{IGBT3} (Ω)	α_{RIGBT1} (K^{-1})	α_{RIGBT2} (K^{-1})	α_{RIGBT3} (K^{-1})	b_2 (A)
Parameter Value	0.443	0.195	0.127	3.61×10^{-3}	2.05×10^{-3}	1.9×10^{-3}	1.2

In Figures 3 and 4 measured (points) and computed (lines) characteristics of the diode (Figure 3) and IGBT (Figure 4) are shown. These characteristics are measured and computed at selected values of ambient temperature T_a.

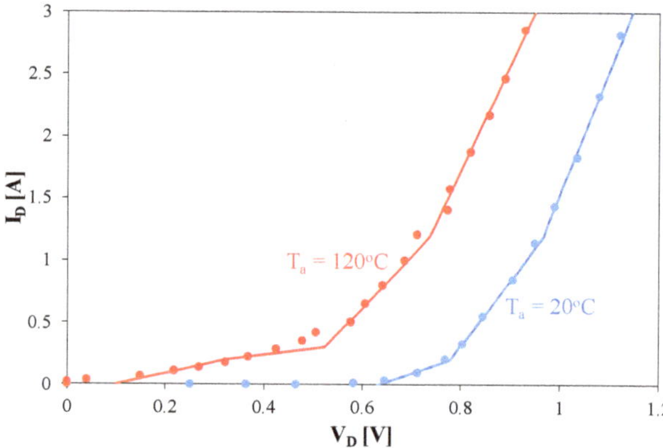

Figure 3. Characteristics of a forward-biased diode computed with the use of a piecewise linear model (lines) and measured (points) at selected values of temperature T_a.

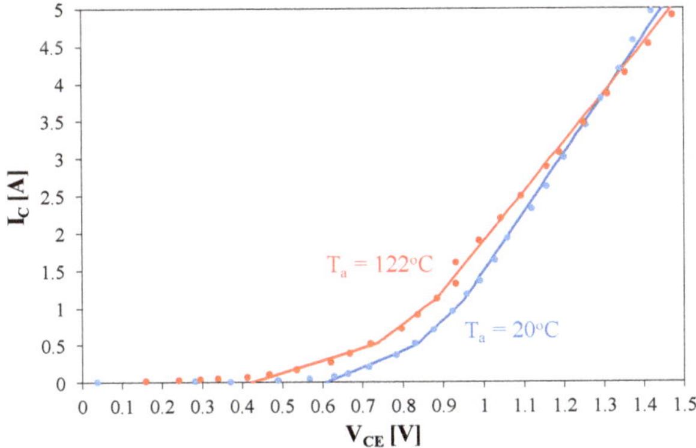

Figure 4. Output characteristics of insulated gate bipolar transistor (IGBT) computed with the use of a piecewise linear model (lines) and measured (points) at selected values of temperature T_a.

As is visible, the results of measurements and computations are convergent for both considered semiconductor devices in a wide range of changes of ambient temperature T_a. As can be observed, all characteristics are described using 4 line segments. A slope of the considered characteristics increases with an increase of the main current of the tested device. This means that series resistances of the diode and IGBT decrease with an increase of this current. It is also observed that the diode forward-voltage and on-voltage of IGBT at zero current decreases with an increase in ambient temperature.

With the diode, an increase in temperature causes a decrease in diode forward-voltage. In turn, for IGBT with a range of low collector current values, an increase in ambient temperature causes a decrease in on–voltage of IGBT. In contrast, for high collector current values, an increase in temperature T_a causes an increase in on–voltage of IGBT.

The presented characteristics correspond to isothermal operating conditions of tested devices, at which self-heating phenomena can be omitted. In the real case, device internal temperature is higher than ambient temperature as a result of the mentioned phenomena [26,27,33]. The cooling conditions of this device are characterized by thermal resistance. This parameter is measured using impulse electrical methods described in [34–36].

Both considered semiconductor devices are situated in the TO-220 case. The measured values of IGBT and the diode are nearly the same and they are equal to about 44 K/W. Reference temperature T_0 is equal to 20 °C.

4. Results of Measurements and Computations

In order to verify usefulness of the proposed electrothermal averaged model of a diode–transistor switch, measurements and computations of characteristics of the boost converter with considered semiconductor devices were performed. A diagram of the tested converter is shown in Figure 5, whereas a photo of the tested converter is presented in Figure 6.

In the considered converter, the input voltage V_{in} was equal to 12 V, and R_0 was load resistance. The inductance of inductor L_1 was equal to 560 µH and capacitance of capacitor C_1 was equal to 1 mF. Voltage source V_{ctrl} produced a rectangular pulsed train of frequency f equal to 10 kHz and duty cycle d. Ammeters were used to measure the input and output currents of the tested converter. Internal resistance of these ammeters was equal to 0.31 Ω. The prototype was mounted on a PCB; the diode of the type IDP08E65 and IGBT of the type IGP06N60T operated without any heat-sinks. The control signal was given by a signal generator exciting gate driver IR2125 by Infineon Technologies.

Some waveforms of voltages and currents of the tested DC–DC converter were measured using an oscilloscope Rigol MS05104 and current probe Tektronix PCPA 300 for different parameters of the control signal and load resistances. For example, in Figure 7 measured waveforms of $v_{GE}(t)$

voltage (yellow line), $v_{CE}(t)$ voltage (violet line) and $i_L(t)$ current (blue line) are shown. The mentioned quantities are marked in Figure 5. Waveform of $i_L(t)$ was obtained after conversion of the measured current into voltage in the current probe. The conversion coefficient was equal to 1 A/V. During these measurements load resistance R_0 was equal to 47 Ω.

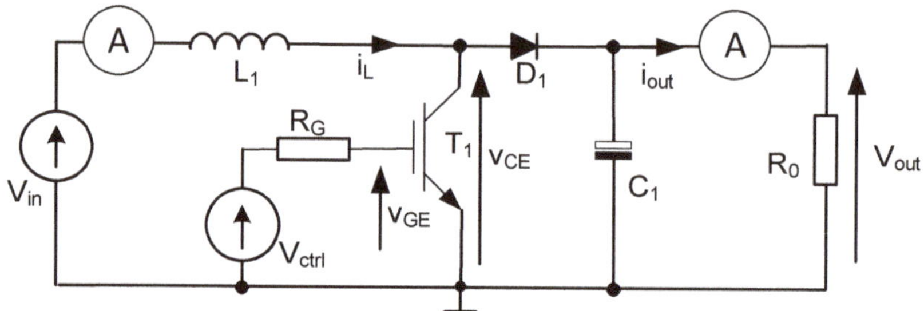

Figure 5. Diagram of the tested boost converter.

Figure 6. Photo of the tested boost converter.

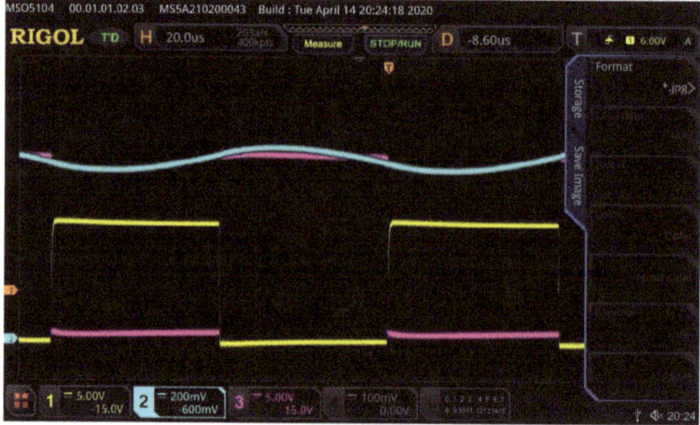

Figure 7. Measured waveforms of $v_{GE}(t)$ voltage, $v_{CE}(t)$ voltages and $i_L(t)$ current in the tested boost converter operating at load resistance R_0 = 47 Ω.

As it can be observed, the signal controlling the input of IGBT (v_{GE} voltage) had frequency equal to 10 kHz, duty cycle equal to 0.5, low level voltage equal to zero and high level voltage equal to 15 V. These parameters values of the control signal were adequate for the used transistor. The transistor output voltage $v_{CE}(t)$ had a shape of a rectangular pulsed train. The highest level of this voltage was equal to about 22 V. The current $i_L(t)$ had positive values only, which proved that the tested converter operated in CCM.

It is important to notice that the tested DC–DC converter operated without any feedback loop, typically used in switch-mode power supplies including such converters. At the chosen operating condition, the influence of parameters of the control signal and on the load resistance were not compensated by the feedback loop. Therefore, any disadvantages of the proposed model can be clearly illustrated for the tested circuit.

Selected resulted of measurements and computations of the considered DC–DC converter are shown in the successive figures. In these figures points denote the results of measurements, solid lines—the results of computations performed with the use of the proposed electrothermal averaged model of a diode–transistor switch including IGBT and a rapid switching diode (called also the new model), black dashed lines—the results of computations performed with the use of the averaged model of a diode transistor switch with ideal switches described e.g., in [5,8] and blue dotted lines—the results of computations performed with the use of the averaged model of a diode–transistor switch including IGBT described in [21].

Figures 8–11 present computed and measured characteristics of the considered converter operating at the fixed value of duty cycle d = 0.5 and the varied value of load resistance R_0. In turn, Figures 12–15 show characteristics of this converter operating at the fixed value of load resistance R_0 = 47 Ω and the varied value of duty cycle d. Values of voltages and currents were measured with the used of laboratory voltmeters and ammeters. The diode and transistor temperatures were measured with the use of an infrared method performed with a pyrometer PT-3S by Optex [37]. This instrument made it possible to measure the case temperature of the mentioned semiconductor devices. Due to a very small value of junction–case thermal resistance of the considered semiconductor devices—which was much lower than junction-ambient thermal resistance—a difference between internal and case temperatures of these devices was not higher than 5 °C.

In Figure 8 measured and computed dependences of converter output voltage V_{out} on load resistance R_0 are presented. As is visible, in the considered operating conditions, the boost converter operated in CCM for R_0 < 100 Ω and in DCM for R_0 > 100 Ω. In both modes of operation, the new model guarantees good accuracy of computations. At small values of load resistance, a decrease was visible in the value of output voltage caused by influence of a voltage drop on the switched on transistor and diode. This voltage drop was an increasing function of converter output current and a decreasing function of load resistance R_0. Literature models described in [5,8,21] could be used for the converter operating in CCM only, because differences between the results of computations performed with these models and the results of measurements were acceptable only in this mode. Of course, the results performed with the use of the model proposed in [21] were more convergent with the results of measurements in CCM than the results performed with the model given in [5,8].

Figure 9 presents dependences of watt-hour efficiency of the considered DC–DC converter on load resistance.

The obtained values of watt-hour efficiency were in the range from 0.8 to 0.9. Values of this parameter computed using the electrothermal model differed from the results of measurements by no more than 7%. The resulted obtained using the considered literature models were overstated even by 17%.

In Figure 10 dependence of internal temperature of IGBT on load resistance was shown. Such dependence could be obtained with the use of the electrothermal model only.

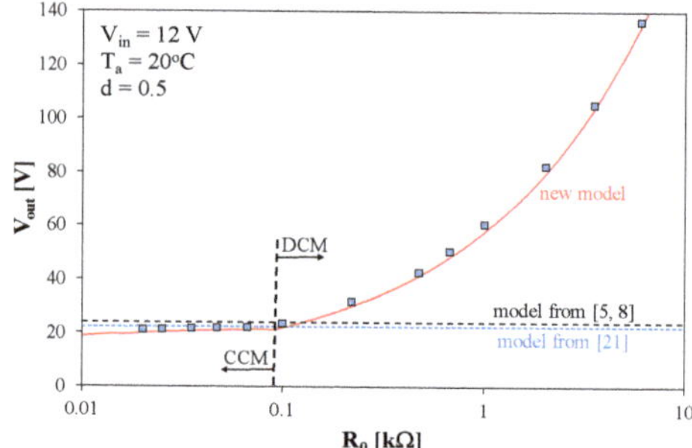

Figure 8. Computed and measured dependences of boost converter output voltage on load resistance.

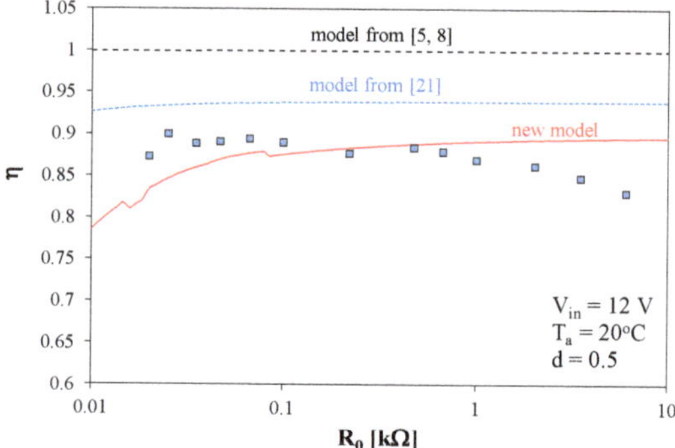

Figure 9. Computed and measured dependences of watt-hour efficiency of the boost converter on load resistance.

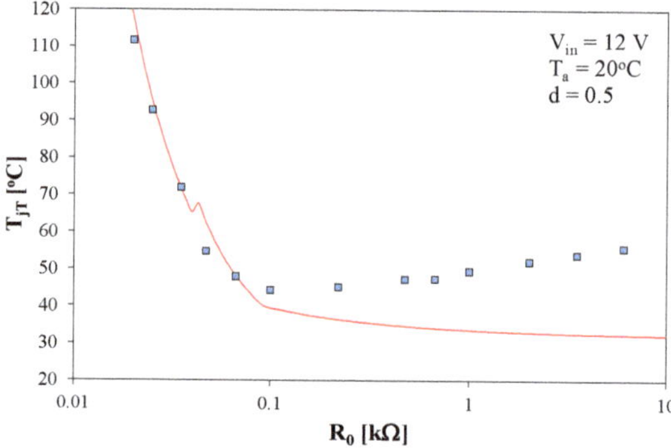

Figure 10. Computed and measured dependence of internal temperature of IGBT on load resistance.

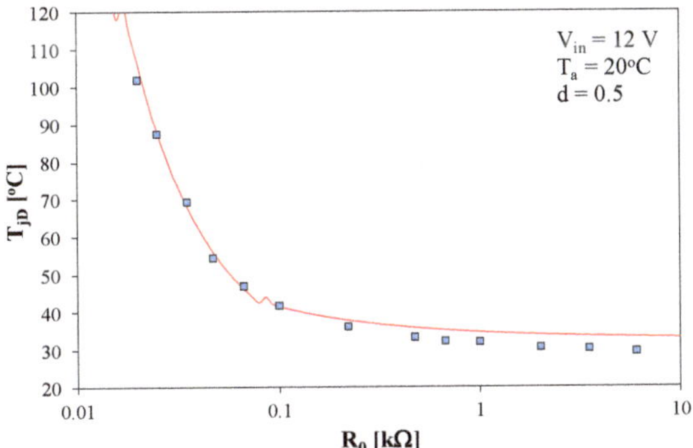

Figure 11. Computed and measured dependence of internal temperature of the diode on load resistance.

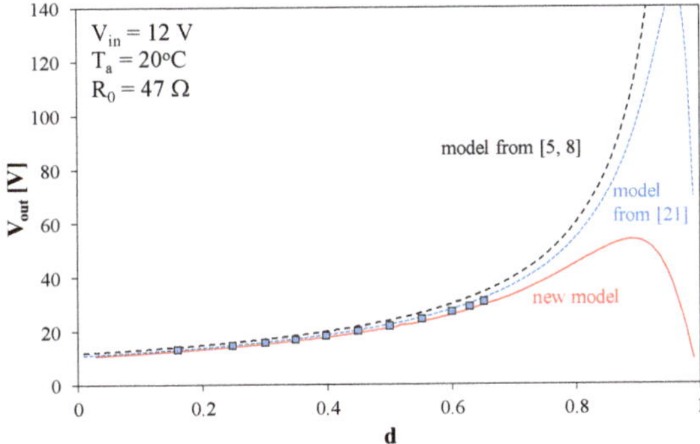

Figure 12. Computed and measured dependences of boost converter output voltage on duty cycle.

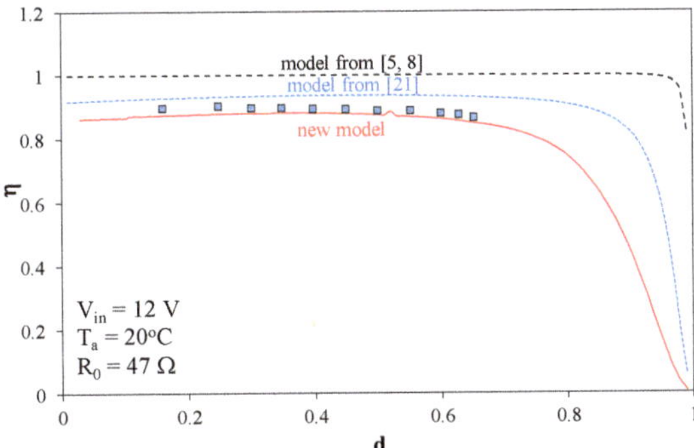

Figure 13. Computed and measured dependences of watt-hour efficiency of boost converter on duty cycle.

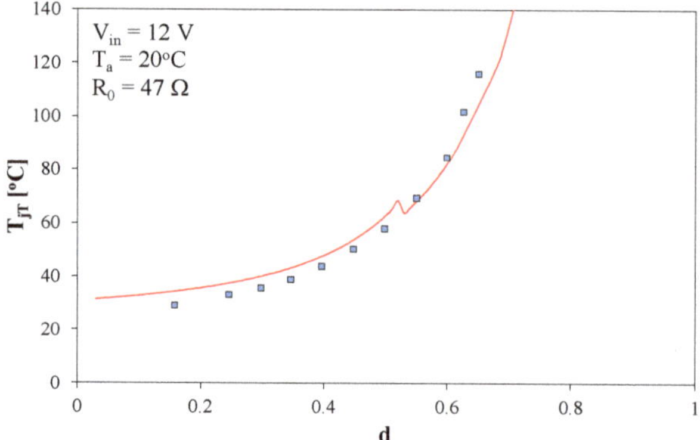

Figure 14. Computed and measured dependences of internal temperature of IGBT on duty cycle.

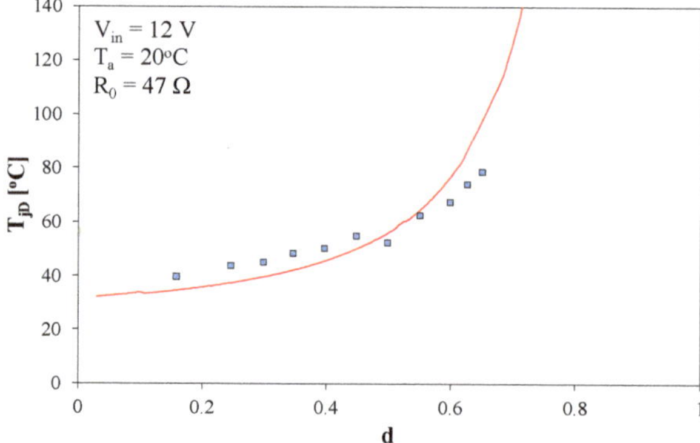

Figure 15. Computed and measured dependences of internal temperature of the diode on duty cycle.

As can be seen, very good accuracy in modeling this dependence was obtained only was CCM. In DCM differences between the results of measurements and computations increased with an increase in load resistance. They exceeded even 20 °C at $R_0 > 6$ kΩ which could be connected with a long switching-off process of IGBT [38], that was not taken into account in the proposed model.

In Figure 11 measured and computed (using the new model) dependence of internal temperature of diode on load resistance is shown.

The considered dependence is a monotonically decreasing function. The result of computations were convergent with the results of measurements. Differences did not exceed 4 °C. Of course, the considered literature models did not make it possible to compute such dependence.

Figure 12 shows computed (with the new model) and measured dependences of converter output voltage on duty cycle.

In the whole considered range of changes of duty cycle the tested boost converter operated in CCM. The measured dependence $V_{out}(d)$ is a monotonically increasing function, but measurements were performed only for $d < 0.67$. The range of changes of d was limited due to a high increase in internal temperature of IGBT, which attained even 120 °C. The considered dependences computed using the electrothermal model and the model given in [21] had maxima at $d > 0.9$. These maxima were observed due to voltage drops on switched-on semiconductor devices [16]. It was also visible that due to self-heating the value of these maxima decrease even three times. Differences between the results of measurements and computations were small and they do not exceed 5%.

Figure 13 illustrates dependence of watt-hour efficiency of the tested boost converter on duty cycle.

As can be observed, dependences η(d) were monotonically decreasing functions. The results of computations performed with the electrothermal model were convergent with the result of measurements. Differences do not exceed 2%. Values of watt-hour efficiency computed using the other models were overstated even by 15%.

Figure 14 presents measured and computed (with the new model) dependence of internal temperature of IGBT on duty cycle.

The considered dependence was an increasing function. The results of computations differ from the results of measurements not more than 5 °C.

Figure 15 shows measured and computed (with the new model) dependence of internal temperature of the diode on duty cycle.

It can be easily observed that temperature T_{jD} increased with an increase in duty cycle. This means that the power dissipated in this diode increased, despite the fact that the time in which this diode was forward-biased decreased. The differences between the results of computations and measurements did not exceed a few degrees Celsius.

Figure 16 illustrates influence of duty-cycle on the output characteristics of the tested DC–DC converter (Figure 16a) and the dependence of diode internal temperature on the converter's output current i_{out}.

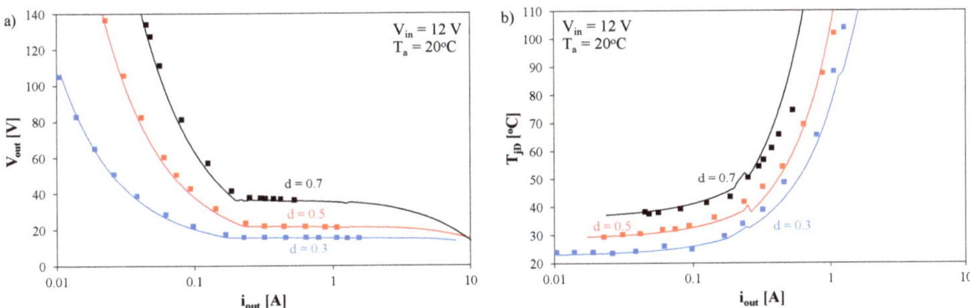

Figure 16. Computed and measured output characteristics of the tested DC–DC converter (**a**) and dependences of internal temperature of the diode on converter output current (**b**) for selected values of duty cycle.

As it can be observed in Figure 16a, very good agreement between computed and measured output characteristics of the tested converter was obtained at all the considered values of duty cycle. The considered dependence was a decreasing function. The critical output current at which the mode of DC–DC converter operation change from DCM to CCM decreased with an increase in duty cycle d. In turn, in Figure 16b, it can be observed that the dependence $T_{jD}(i_{out})$ was an increasing function and values of temperature T_{jD} increase with an increase in duty cycle.

An influence of ambient temperature on characteristics of the tested boost converter is illustrated in Figure 17.

It is clearly visible in Figure 17a that changes in value of ambient temperature practically do not influence value of output voltage of the tested boost converter. Points marking results of measurements performed in both the values of ambient temperature practically overlap as well as lines representing computations performed for these values of T_a. Changes in value of V_{out} voltage caused by change in ambient temperature did not exceed 0.3 V. Hence, small value of these changes was a result of weak influence of temperature on IGBT output voltage. The same result were obtained during computations and measurements. In contrast, in Figure 17b, it is visible that change in the value of ambient temperature caused practically the same change in the value of internal temperature of the diode. The same influence of ambient temperature the Authors observed also for internal temperature of IGBT.

Some additional computations of the tested boost converter were performed using the worked-out electrothermal averaged model of a diode transistor switch, including the IGBT and a rapid switching diode. For example, Figure 18 illustrated influence of load resistance on dependences of converter output voltage (Figure 18a) and internal temperature of the diode (Figure 18b) on the duty cycle.

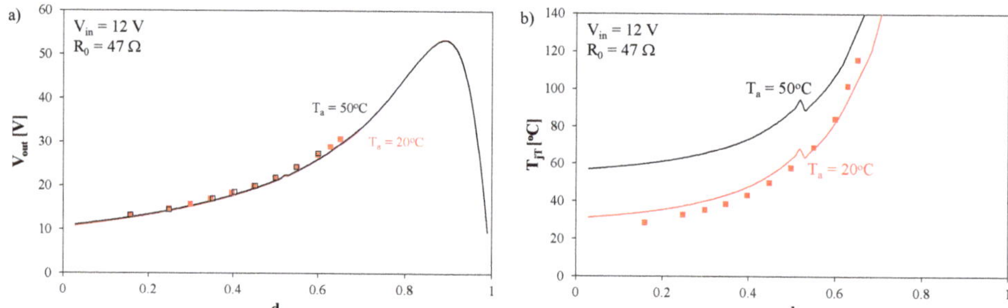

Figure 17. Computed and measured dependences of converter output voltage (**a**) and internal temperature of the diode (**b**) on duty cycle.

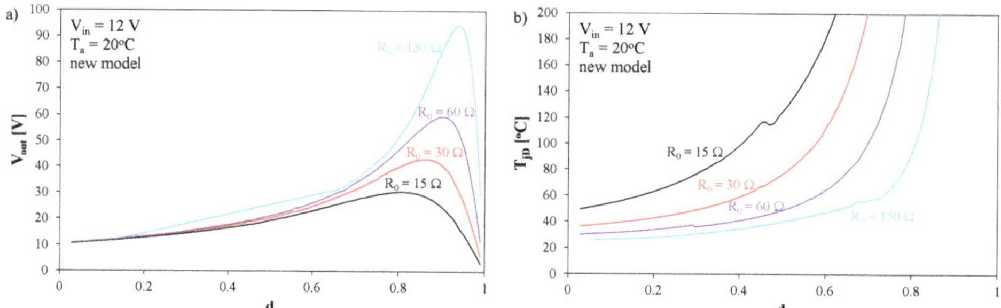

Figure 18. Computed dependences of converter output voltage (**a**) and internal temperature of the diode (**b**) on duty cycle for selected values of load resistance.

As it can be observed, an increase of load resistance R_0 caused an increase in the maximum value of converter output voltage and an increase in the value of duty cycle, at which the maximum in characteristic $V_{out}(d)$ was observed. The maximum value of V_{out} voltage changes in the range from 30 to 94 V. In turn, internal temperature of the diode was an increasing function of duty cycle and a decreasing function of load resistance. It is worth noticing that at the considered cooling conditions of the diode the acceptable value of duty cycle was strongly limited. For the considered values of R_0 the maximum value of duty cycle increased from about 0.5 (for $R_0 = 15\ \Omega$) to about 0.8 (for $R_0 = 150\ \Omega$).

Figure 19 illustrates the influence of duty cycle on dependences of converter output voltage (Figure 19a) and internal temperature of IGBT (Figure 19b) on load resistance.

In Figure 19a it is visible that dependence $V_{out}(R_0)$ was an increasing function for all considered values of duty cycle d. It could also be observed that the value of load resistance at which the border between the CCM and DCM exists moves right with an increase in the value of d. In CCM influence of energy losses in semiconductor devices on converter output voltage were much more visible for the highest value of duty cycle. In turn, in Figure 19b it can be seen that internal temperature of IGBT is a decreasing function of load resistance and an increasing function of duty cycle. Cooling conditions of IGBT limit the lowest admissible value of load resistance at the adjusted duty cycle. In the considered case for d = 0.8 load resistance should not be smaller than 100 Ω in order to limit the value of internal temperature of IGBT to 150 °C.

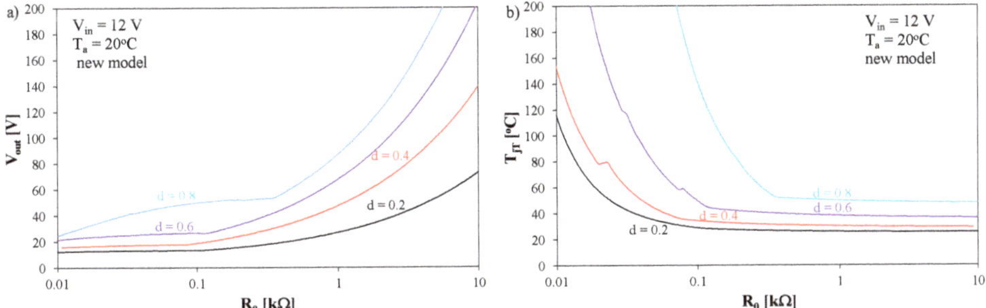

Figure 19. Computed dependences of converter output voltage (**a**) and internal temperature of IGBT (**b**) on load resistance for selected values of duty cycle.

5. Conclusions

In this study, a new electrothermal averaged model of the diode transistor switch—including IGBT and a rapid switching diode—is proposed. This model makes it possible to compute non-isothermal (taking into account self-heating phenomena) characteristics of DC–DC converters operating both in CCM and DCM. Values of voltages, currents and internal temperatures of semiconductor devices at the thermally steady state can be computed using this model. The proposed model uses piecewise linear approximation of current–voltage characteristics of IGBT and the diode operating in on–state.

Practical usefulness of the proposed model was verified for the boost converter operating both in CCM and DCM at different values of duty cycle of the control signal in a wide range of load resistance. It was shown that the new electrothermal averaged model makes it possible to obtain good accuracy of modeling dependences $V_{out}(d)$, $V_{out}(R_0)$, $\eta(R_0)$, $\eta(d)$ and $V_{out}(i_{out})$ in both operating modes of the converter. This accuracy is much higher than accuracy obtained with the use of the considered literature models, particularly in DCM.

The proposed model makes it also possible to compute values of internal temperature of both semiconductor devices operating in the tested DC–DC converter. The obtained results of computations of internal temperature of the diode are convergent with the results of measurements in all considered modes of operation of the DC–DC converter. In contrast, the results of computations of internal temperature of the IGBT fit well the results of measurements only for the DC–DC converter operating in CCM. In DCM, these differences can exceed even 20 °C. It was also shown that changes in the value of ambient temperature practically do not influence on the DC–DC converter operating at fixed values of duty cycle and load resistance. These changes cause changes in internal temperatures of IGBT and the diode. These changes are very important from the point of view of estimation life time of the tested converter. As it is known from the literature [39], an increase in value of device internal temperature by 8 °C causes even twice decrease in life time of such devices.

Computations performed with the use of the proposed model illustrate influence of load resistance and duty-cycle of the control signal on selected characteristics of the boost converter. The obtained results of computations make it possible to estimate the range of admissible values of the mentioned parameters, for which internal temperature of the diode and IGBT does not exceed the admissible value given by the producer.

In further investigations, the authors will attempt to find a cause of these differences and will propose some modifications of the model. The model will be also tested in a wide range of frequency of the control signal and for different cooling conditions of semiconductor devices operating in the DC–DC converter, as well as at different values of input voltage, load current and ambient temperature.

Author Contributions: Conceptualization (P.G. and K.G.); computations (P.G. and K.G.), methodology (P.G. and K.G.); experimental verification (P.G.); writing text of the article (K.G. and P.G.); review and editing (K.G. and P.G.); visualization (P.G. and K.G.); supervision (K.G.). All authors have read and agreed to the published version of the manuscript.

Funding: The scientific work is a result of the project No. 2018/31/N/ST7/01,818 financed by the Polish National Science Center.

Conflicts of Interest: The authors declare no conflicts of interest.

References

1. Rashid, M.H. *Power Electronic Handbook*; Academic Press: Cambridge, MA, USA; Elsevier: Amsterdam, The Netherlands, 2007.
2. Kazimierczuk, M. *Pulse-Width Modulated DC-DC Power Converters*; Wiley: Hoboken, NJ, USA, 2015.
3. Billings, K.; Morey, T. *Switch-Mode Power Supply Handbook*; McGraw-Hill Companies: New York, NY, USA, 2011.
4. Ang, S.; Oliva, A. *Power—Switching Converters*; CRC Press Taylor and Francis Group: Boca Raton, FL, USA, 2011.
5. Ericson, R.; Maksimovic, D. *Fundamentals of Power Electronics*; Kluwer Academic Publisher: Norwell, MA, USA, 2001.
6. Rashid, M.H. *Spice for Power Electronics and Electric Power*; CRC Press: Boca Raton, FL, USA, 2006.
7. Wells, B.A.; Brodlo, B.T.; Babaa, I.M.H. Analog computer simulation of a Dc-to-Dc flyback converter. *IEEE Trans. Aerosp. Electron. Syst.* **1967**, *AES-3*, 399–409.
8. Basso, C. *Switch-Mode Power Supply SPICE Cookbook*; McGraw-Hill: New York, NY, USA, 2001.
9. Vladirmirescu, A. Shaping the History of SPICE. *IEEE Solid-State Circuits Mag.* **2011**, *3*, 36–39.
10. Maksimovic, D.; Stankovic, A.M.; Thottuvelil, V.J.; Verghese, G.C. Modeling and simulation of power electronics converters. *Proc. IEEE* **2001**, *89*, 898–912. [CrossRef]
11. Górecki, K.; Zarębski, J. The Method of a Fast Electrothermal Transient Analysis of Single-Inductance DC-DC Converters. *IEEE Trans. Power Electron.* **2012**, *27*, 4005–4012. [CrossRef]
12. Vorperian, V. *Fast Analitycal Techniques for Electrical and Electronic Circuits*; Cambridge University Press: Cambridge, UK, 2002.
13. Bedrosian, D.; Vlach, J. Time-domain analysis of networks with internally controlled switches. *IEEE Trans. Circuits Syst. I Fund. Theory Appl.* **1992**, *39*, 199–212. [CrossRef]
14. Ben-Yaakov, S.; Gaaton, Z. Generic SPICE compatible model of current feedback in switch mode convertors. *Electron. Lett.* **1992**, *28*, 1356–1358. [CrossRef]
15. Ben-Yaakov, S.; Glozman, S.; Rabinovici, R. Envelope simulation by SPICE-compatible models of electric circuits driven by modulated signals. *IEEE Trans. Ind. Electron.* **2000**, *47*, 222–225. [CrossRef]
16. Górecki, K. A new electrothermal average model of the diode-transistor switch. *Microelectron. Reliab.* **2008**, *48*, 51–58. [CrossRef]
17. Górecki, K.; Detka, K. Application of Average Electrothermal Models in the SPICE-Aided Analysis of Boost Converters. *IEEE Trans. Ind. Electron.* **2019**, *66*, 2746–2755. [CrossRef]
18. Pavlovic, T.; Bjazic, T.; Ban, Z. Simplified Averaged Models of DC-DC Power Converters Suitable for Controller Design and Microgrid Simulation. *IEEE Trans. Power Electron.* **2013**, *28*, 3266–3275. [CrossRef]
19. Meo, S.; Toscano, L. Some New Results on the Averaging Theory Approach for the Analysis of Power Electronic Converters. *IEEE Trans. Ind. Electron.* **2018**, *65*, 9367–9377. [CrossRef]
20. Han, J.; Zhang, B.; Qiu, D. Bi-switching Status Modeling Method for DC-DC Converters in CCM and DCM Operations. *IEEE Trans. Power Electron.* **2017**, *21*, 2464–2472. [CrossRef]
21. Górecki, P. Application of the averaged model of the diode-transistor switch for modelling characteristics of a boost converter with an IGBT. *Int. J. Electron. Telecommun.* **2020**. in review.
22. Bryant, A.; Allotey, N.A.P.; Hamilton, D.; Swan, I.; Mawby, P.; Ueta, T.; Nishijima, T.; Hamada, K. A Fast Loss and Temperature Simulation Method for Power Converters, Part I: Electrothermal Modeling and Validation. *IEEE Trans. Power Electron.* **2012**, *27*, 248–257. [CrossRef]
23. Górecki, K.; Zarębski, J. Modeling nonisothermal characteristics of switch-mode voltage regulators. *IEEE Trans. Power Electron.* **2008**, *23*, 1848–1858. [CrossRef]
24. Górecki, K. Non-linear average electrothermal models of buck and boost converters for SPICE. *Microelectron. Reliab.* **2009**, *49*, 431–437. [CrossRef]
25. Bryant, B.; Kazimierczuk, M.K. Voltage-Loop Power-Stage Transfer Functions with MOSFET Delay for Boost PWM Converter Operating in CCM. *IEEE Trans. Ind. Electron.* **2007**, *54*, 347–353. [CrossRef]

26. Starzak, Ł.; Zubert, M.; Janicki, M.; Torzewicz, T.; Napieralska, M.; Jabłoński, G.; Napieralski, A. Behavioral approach to SiC MPS diode electrothermal model generation. *IEEE Trans. Electron Devices* **2013**, *60*, 630–638. [CrossRef]

27. Mawby, P.A.; Igic, P.M.; Towers, M.S. Physically based compact device models for circuit modelling applications. *Microelectron. J.* **2001**, *32*, 433–447. [CrossRef]

28. Wilamowski, B.M.; Jaeger, R.C. *Computerized Circuit Analysis Using SPICE Programs*; McGraw-Hill: New York, NY, USA, 1997.

29. IGP06N60T, Datasheet, Infineon Technologies. Available online: https://www.infineon.com/dgdl/Infineon-IGP06N60T-DS-v02_03-en.pdf?fileId=db3a30432313ff5e0123b82d13ba7883 (accessed on 7 April 2020).

30. IDP08E65D1, Datasheet, Infineon Technologies. Available online: https://www.alldatasheet.com/datasheet-pdf/pdf/756371/INFINEON/IDP08E65D1.html (accessed on 7 April 2020).

31. Measuring HEXFET Characteristics, Application Note AN-957, International Rectifier. Available online: https://www.infineon.com/dgdl/an-957.pdf?fileId=5546d462533600a40153559f0dfc11dc (accessed on 7 April 2020).

32. Keithley 2612a, Datasheet. Available online: http://www.testequipmenthq.com/datasheets/KEITHLEY-2612A-Datasheet.pdf (accessed on 11 June 2020).

33. Górecki, P.; Górecki, K. Influence of thermal phenomena on dc characteristics of the IGBT. *Int. J. Electron. Telecommun.* **2018**, *64*, 71–76.

34. Górecki, K.; Górecki, P.; Zarębski, J. Measurements of parameters of the thermal model of the IGBT module. *IEEE Trans. Instrum. Meas.* **2019**, *68*, 4864–4875. [CrossRef]

35. Blackburn, D.L. Temperature Measurements of Semiconductor Devices—A Review. In Proceedings of the 20th IEEE Semiconductor Thermal Measurement and Menagement Symposium SEMI-THERM, San Jose, CA, USA, 11 March 2004; pp. 70–80.

36. Górecki, K.; Zarębski, J. Modeling the influence of selected factors on thermal resistance of semiconductor devices. *IEEE Trans. Compon. Packag. Manuf. Technol.* **2014**, *4*, 421–428. [CrossRef]

37. Pyrometer PT-3S, Optex Product Information. Available online: http://www.dewetron.cz/optex/katlisty/PT-3S.pdf (accessed on 11 June 2020).

38. Górecki, P.; Górecki, K. Modelling a Switching Process of IGBTs with Influence of Temperature Taken into Account. *Energies* **2019**, *12*, 1894. [CrossRef]

39. Narendran, N.; Gu, Y. Life of LED-based white light sources. *J. Disp. Technol.* **2005**, *1*, 167–171. [CrossRef]

 *energies*

Article

Electrothermal Model of SiC Power BJT

Joanna Patrzyk *, Damian Bisewski and Janusz Zarębski

Department of Marine Electronics, Gdynia Maritime University, 81-225 Gdynia, Poland;
d.bisewski@we.umg.edu.pl (D.B.); j.zarebski@we.umg.edu.pl (J.Z.)
* Correspondence: j.patrzyk@we.umg.edu.pl; Tel.: +48-58-5586-575

Received: 28 March 2020; Accepted: 17 May 2020; Published: 21 May 2020

Abstract: This paper refers to the issue of modelling characteristics of SiC power bipolar junction transistor (BJT), including the self-heating phenomenon. The electrothermal model of the tested device is demonstrated and experimentally verified. The electrical model is based on the isothermal Gummel–Poon model, but several modifications were made including the improved current gain factor (β) model and the modified model of the quasi-saturation region. The accuracy of the presented model was assessed by comparison of measurement and simulation results of selected characteristics of the BT1206-AC SiC BJT manufactured by TranSiC. In this paper, a single device characterization has only been performed. The demonstrated results of research show the evident temperature impact on the transistor d.c. characteristics. A good compliance between the measured and calculated characteristics of the considered transistor is observed even in quasi-saturation mode.

Keywords: BJT; modelling; self-heating; silicon carbide; SPICE

1. Introduction

While the power MOSFET made of silicon carbide (SiC) is currently the most popular switching power wide bandgap-based semiconductor device, the SiC bipolar junction transistor (BJT) is still an inviting counterpart dedicated to high-temperature/high-power systems, due to its particular properties, such as low specific on-resistance, a less complicated manufacture method and its being free of oxide reliability referred to the high value of the electrical field and temperature. Since the last decade, the market has been offering high voltage silicon carbide power BJTs [1–5].

The research subject in this paper is the high voltage BJT made of silicon carbide, BT1206-AC manufactured by the TranSiC company and placed in a typical TO-247 case. The values of the certain parameters of this transistor are demonstrated in Table 1 [5].

Table 1. Parameter values of the BT1206-AC transistor.

Manufacturer	Type	V_{CEO} [V]	I_{Cmax} [A]	T_{jmax} [°C]	β [A/A]	Package
TranSiC	BT1206-AC	1200	6	175	35 at (V_{CE} = 2.5 V, I_C = 4 A)	TO-247

Due to the improved electrical and thermal properties of silicon carbide devices compared to silicon devices, companies state that designers will want to replace silicon components for applications such as hybrid electric vehicles, inverters of solar cells and high power electric motor controllers with their silicon carbide counterparts. For all these applications, silicon carbide can solve the problems with a high-tempergature environment and switching losses.

The ambient temperature has a strong impact on certain characteristics and parameters of the silicon carbide power bipolar transistors [4,6–8]. Moreover, in real operating conditions, due to the self-heating phenomenon, the junction temperature exceeds the ambient temperature because of the device dissipated thermal power converted into the heat in non-ideal heat dissipation conditions. The self-heating phenomenon, resulting in qualitative and quantitative changes of the semiconductor device characteristics shape, is observed as well in SiC BJTs [6].

In order to properly analyze the operation of SiC BJT-based electronic systems, a precise and relatively simple SiC BJT model for circuit simulation and trend analysis is needed. One of the earliest semi-physical models for silicon BJT is the Gummel and Poon model (G–P model) proposed in 1970 [9]. The G–P model has become a fundamental model for characteristics calculation of BJT in SPICE (Simulation Program with Integrated Circuit Emphasis) which is a popular computer tool that allows the analysis of characteristics and parameters of electronic components and systems [7,10–14]. Certain modifications have been made to evolve SiC BJT models in recent years. In 2004, the G–P model was applied to calculate the characteristics of SiC BJT for the first time [15]. Johannesson and Nawaz established a G–P-based analytical SPICE model for SiC BJT power modules that is useful for the designing of power electronics systems containing considered devices [16]. However, these models do not take into account the self-heating phenomenon. There are some electrothermal models (ETMs) dedicated to SiC BJT transistors in the literature [6,17,18], but works describing these models do not contain full information enabling their implementation in the SPICE program (e.g., no description of the controlled sources used in the model) or use numerical models that are complex, demand specific data concerning material features or device geometry and have time-consuming simulations.

This paper refers to the issue of modelling of static characteristics of SiC power BJTs including the self-heating phenomenon. The electrothermal model of SiC BJT has been proposed. The model was experimentally examined by comparison of the measured and simulated isothermal and non-isothermal characteristics of the selected transistor. The effect of self-heating on transistor characteristics was assessed and discussed. In this paper, the experimental verification of the model has been performed using a single device characterization, however, future work will include verification results based on a larger population of the considered class of devices.

2. The Model Form

A block diagram of the electrothermal model (ETM) of SiC BJT proposed by the authors is demonstrated in Figure 1.

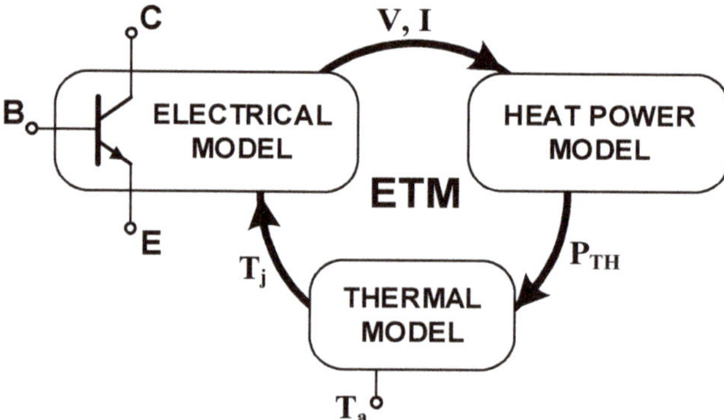

Figure 1. A block diagram of the electrothermal model of SiC bipolar junction transistor (BJT).

As presented, the ETM contains: the electrical model, the heat power model and the thermal model. The function of the electrical model is to determine the static characteristics of the transistor for a known temperature (T_j) of its interior. The heat power model is designed to determine the value of electrical power generated in the transistor based on the currents and clamping voltages values on the transistor leads. The thermal model calculates the actual value of transistor junction (inner) temperature based on the ambient temperature (T_a) and the thermal parameters associated with the transistor cooling method.

The form of the considered electrical model is based on the G–P model. However, in our model, a voltage-dependent collector resistance is used for modelling the quasi-saturation mode that starts to be visible in the higher temperature range in SiC BJTs' static output characteristics. The quasi-saturation region model is based on the Kull model and consists of the temperature-dependent intrinsic carrier concentration model dedicated to the devices made of silicon carbide. A network form of the proposed electrical model is demonstrated in Figure 2 [1,16,19].

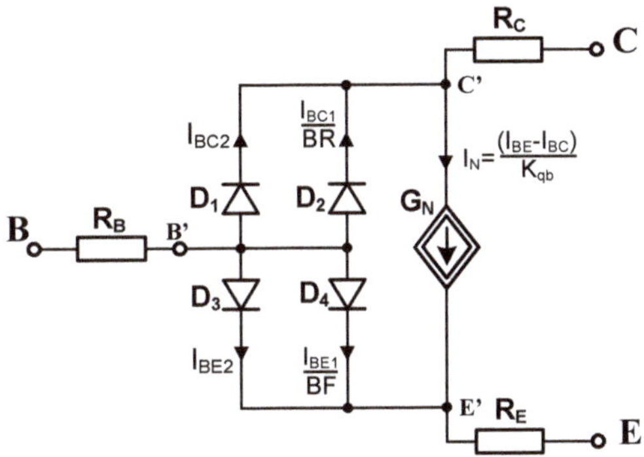

Figure 2. Network form of the electrical model of SiC BJT.

Diodes D_1–D_4 model the generation and recombination phenomena in the layers of the collector/emitter-base junctions, whereas D_1–D_2 and D_3–D_4 represent the ideal and non-ideal component of these junctions, respectively. Values of the collector and the base current of the transistor are of the form [19]:

$$I_B = \frac{I_{BE1} + I_{BC1}}{\beta} + I_{BE2} + I_{BC2} \tag{1}$$

$$I_C = \frac{I_{BE1}}{K_{qb}} - \frac{I_{BC1}}{K_{qb}} - \frac{I_{BC1}}{\beta} - I_{BC2} \tag{2}$$

where the currents I_{BE1}, I_{BE2}, I_{BC1}, I_{BC2} are defined as follows [19]:

$$I_{BE1} = IS \cdot \left[\exp\left(\frac{V_{BE}}{NF \cdot V_t} \right) - 1 \right] \tag{3}$$

$$I_{BE2} = ISE \cdot \left[\exp\left(\frac{V_{BE}}{NE \cdot V_t} \right) - 1 \right] \tag{4}$$

$$I_{BC1} = IS \cdot \left[\exp\left(\frac{V_{BC}}{NR \cdot V_t} \right) - 1 \right] \tag{5}$$

$$I_{BC2} = ISC \cdot \left[\exp\left(\frac{V_{BC}}{NC \cdot V_t} \right) - 1 \right] \tag{6}$$

where β is the current gain factor, V_{BE} is the intrinsic (B'–E') base-emitter voltage, V_{BC} is the intrinsic (B'–C') base-collector voltage, IS, ISE and ISC are the saturation currents, NF and NR are the current emission coefficients, NE and NC are the leakage emission coefficients and coefficient K_{qb} is of the form [19]:

$$K_{qb} = 0.5 \cdot \left(1 - \frac{V_{BC}}{V_{AF}} - \frac{V_{BC}}{V_{AF}} \right)^{-1} \cdot \left[1 + \left(1 + 4 \cdot \left(\frac{I_{BE1}}{IKF} + \frac{I_{BC1}}{IKR} \right) \right)^{NK} \right] \tag{7}$$

where V_{AF} and V_{AR} are early voltages, IKF and IKR are the knee current parameters, and NK is the high-current roll-off coefficient.

Efficiency of the controlled-current source G_N from Figure 2 is of the form [19]:

$$I_N = \frac{I_{BE} - I_{BC}}{K_{qb}} \tag{8}$$

The coefficient β existing in Equations (1) and (2) are defined as [20]:

$$\beta = \frac{\beta_0 \cdot \left(1 + a \cdot \left(T_j - T_0\right)\right)}{1 + b \cdot \left(1 + c \cdot \left(T_j - T_0\right)\right) \cdot \sqrt{I_C^2 + d_1}} \cdot \left[1 - \sum_i \gamma_i \cdot \exp\left(\alpha_i \cdot \sqrt{I_C^2 + d_1}\right)\right] \tag{9}$$

where T_0 is the nominal temperature, β_0, α_i, γ_i, a, b, c, and d_1 are the model parameters.

Resistors R_E and R_B represent series resistances of the emitter and base areas according to Equation [19]:

$$R_E(T) = R_E \cdot \left(1 + TRE1 \cdot \left(T_j - T_0\right) + TRE2 \cdot \left(T_j - T_0\right)^2\right) \tag{10}$$

$$R_B(T) = R_B \cdot \left(1 + TRB1 \cdot \left(T_j - T_0\right) + TRB2 \cdot \left(T_j - T_0\right)^2\right) \tag{11}$$

where TRE1, TRB1, TRE1 and TRE2 represent linear and quadratic temperature coefficients of the emitter and the base resistances, respectively.

The voltage-dependent collector resistance (R_C) dedicated for the quasi-saturation region modeling is described as [1,16]:

$$R_C = R_{C0} \cdot \left(\frac{1}{2} + \frac{1}{4} \cdot \sqrt{1 + \frac{4 \cdot n_i^2}{N_{epi}^2} \cdot \exp\left(\frac{q \cdot V_{B'C'}}{k \cdot T_j}\right)} + \frac{1}{4} \cdot \sqrt{1 + \frac{4 \cdot n_i^2}{N_{epi}^2} \cdot \exp\left(\frac{q \cdot V_{B'C}}{k \cdot T_j}\right)}\right)^{-1} \tag{12}$$

where N_{epi} is an epi-layer collector doping concentration, $V_{B'C'}$ and $V_{B'C}$ are voltages in the selected end of the collector resistance, q is the elementary charge, and n_i is the intrinsic carrier concentration defined as [1,16]:

$$n_i = A \cdot T_j^{\frac{3}{2}} \cdot \exp\left(-\frac{E_{g0}}{2 \cdot k \cdot T_j}\right) \tag{13}$$

where E_{g0} is SiC bandgap at zero temperature, k is Boltzmann's constant, A is the material constant independent of temperature.

The zero-bias collector resistance R_{C0} existing in Equation (12) is of the form [1,16]:

$$R_{C0} = 2.143 \cdot 10^{-21} \cdot \exp\left(0.1302 \cdot T_j\right) \tag{14}$$

A detailed specification of the mathematical description of the considered model is achievable in the SPICE user manual [19] and other papers [1,16,20].

In the thermal model, the heat interactions between the device interior and the ambient temperature were taken into account. As a rule, the compact thermal model of a semiconductor device is represented by the form of the RC network. Based on this form from other works [21,22], the non-linear compact thermal model of the SiC BJT was demonstrated. Thermal capacitances hardly depend on the temperature in contrast to the thermal resistance that changes with the temperature. In the presented model controlled voltage sources were used only instead of R elements. The network representation of the considered model is shown in Figure 3.

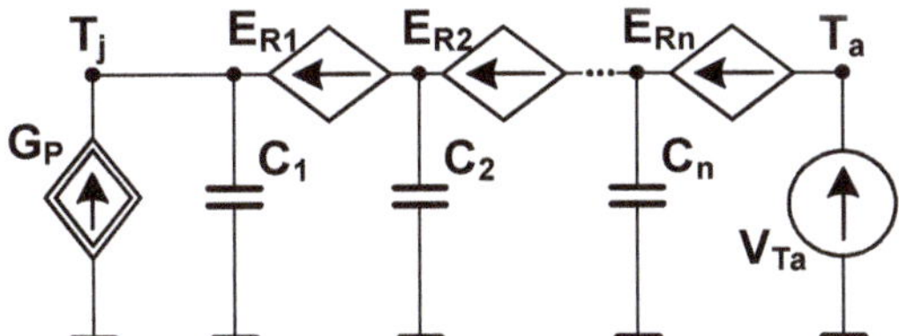

Figure 3. Network form of the non-linear thermal model of the SiC BJT.

The controlled-current source G_P expresses the power dissipated in the device, controlled voltage sources E_{R1}, E_{R2}, ... , E_{Rn} express temperature changes between every part of the thermal dissipation path. The analytical form of these voltage source efficiencies takes into consideration changes in the thermal resistance value expressing the thermal dissipation between the specific construction parts. Capacitors C_1–C_n describe thermal capacitances of every construction part. The voltage at nodes T_j and T_a represents the transistor junction temperature and the ambient temperature respectively.

Efficiencies of controlled-current source G_P and controlled-voltage sources are of the form [21,22]:

$$E_{Rn} = \left(R_{th1} \cdot \exp\left(\frac{i_n}{P_0}\right) + R_{th0}\right) \cdot i_n \cdot d_i \tag{15}$$

$$G_p = P_{th} = I_B \cdot V_{BE} + I_C \cdot V_{CE} \tag{16}$$

where d_i is the quotient of R_n thermal resistance and the transistor thermal resistance R_{th}, R_{th0}, R_{th1}, P_0 are the model parameters, and i_n is the current of the source E_{Rn}.

3. Results of Measurements and Simulations

To test the accuracy of the presented electrothermal model, static isothermal and non-isothermal characteristics of the BT1206-AC SiC BJT were measured and compared to the results of SPICE calculations. The model was implemented in SPICE in the form of the subcircuit. A set of electrical parameter values of the ETM was used in the simulations. The values of the model parameters were acquired from the measured characteristics with the use of the proper parameters estimation procedure described in [20,23]. The values of parameters occurring in Equation (15) were acquired from measurements of the transistor transient thermal impedance.

Isothermal characteristics of the selected transistor were measured by means of Keithley Source Measure Unit type 2602 in order to minimize the impact of the self-heating phenomenon on the measured characteristics of the tested device. During the measurements, a signal with a duty cycle of less than 0.0001 and a pulse duration of 50 µs was used. The non-isothermal characteristics were performed by the "point by point" measurement method for the transistor operating without any cooling system under the thermal steady-state. During the measurements, the case temperature T_C of the transistor was measured by a TM-2000 precision thermometer comprising a platinum Pt-100 sensor. The temperature sensor was glued on the transistor case by a heat conductive AG Termoglue, 10 G.

The measured and simulated isothermal static output characteristics of the BT1206-AC transistor are presented in Figure 4, where points represent the results of measurements and lines represent the results of simulations using the proposed ETM.

As seen in Figure 4, good agreement between the results of the isothermal measurements in comparison with the simulations was obtained even in the quasi-saturation region at both values of the ambient temperature, which expresses a percentage relative error (absolute error quotient and exact value) of less than 8%. The increase of the ambient temperature causes the collector current value drop in the bipolar silicon carbide transistors, but it is commonly known from the literature that in the case of the bipolar transistors made of silicon, this dependence is reversed [24,25].

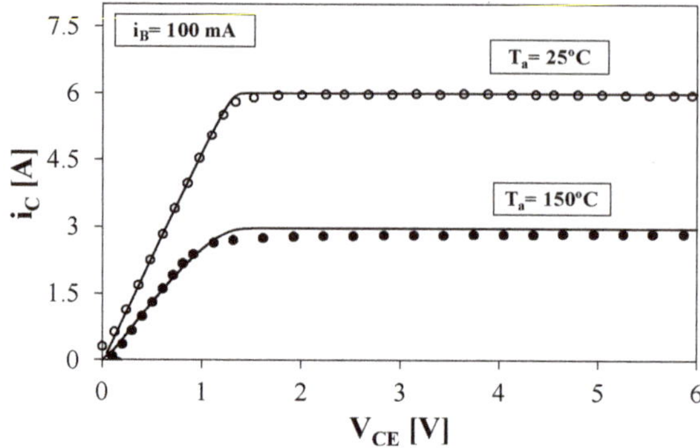

Figure 4. Isothermal output characteristics of the BT1206-AC transistor.

Measured and calculated non-isothermal I-V characteristics of the considered SiC BJT transistor are shown in Figure 5. The investigations were performed at four fixed values of the control current in the range of 10–150 mA. As shown in Figures 4 and 5, the differences between the results of isothermal and non-isothermal measurements of the static output characteristics $i_C(v_{CE})$ reach over 50% for a control current equal to i_B = 100 mA and a voltage v_{CE} of 4 V at room temperature.

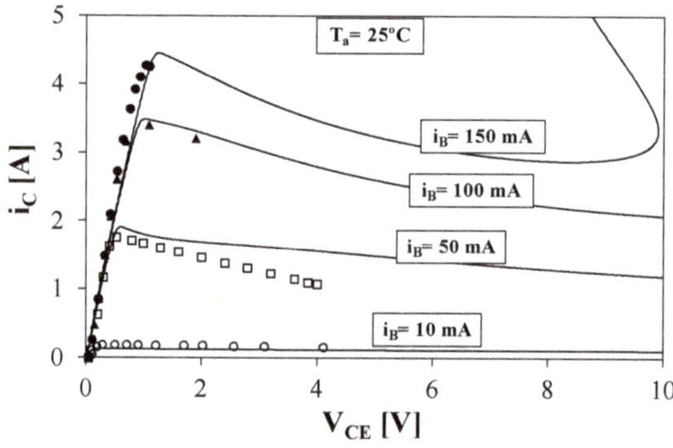

Figure 5. Non-isothermal output characteristics of the BT1206-AC transistor.

As seen, the satisfying agreement between the results of simulations and measurements was expressed by a relative error value of usually less than 10%. The measurements were carried out to obtain the transistor junction temperature (T_j) close to the maximum value of this parameter given in the catalogue by the manufacturer. The junction temperature was estimated on the basis of case temperature (T_C) measurements and catalog value of the thermal resistance R_{thjc} between the transistor junction and case.

The simulated output characteristic at the base current equal to 150 mA has a specific shape. The negative slope of the collector current with increasing collector-emitter voltage is caused by the strong self-heating phenomenon. In the literature, this type of characteristic is called an N-shape characteristic [8].

The important parameter in the description of the bipolar transistors is the common-emitter current gain factor (β), defined as the ratio of the output and control current ($\beta = I_C/I_B$). It is known from the literature that the value of β depends on the operating point. In catalogs, usually only one value of this parameter is given (under set operating conditions). Manufacturers deliberately choose a specific transistor operating point to present the highest value of the current gain factor β. Figure 6 shows the authors' measured and calculated values of β factor for the considered transistor in the isothermal conditions at a fixed value of the collector-emitter voltage equal to 3 V.

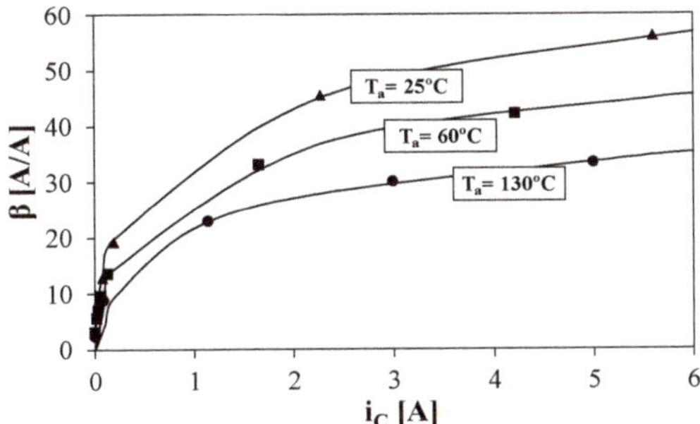

Figure 6. Isothermal dependence of the current gain factor at fixed values of an ambient temperature.

As seen, in the range of collector current values up to about 6 A, the $\beta(i_C)$ dependence is a monotonically increasing function. In addition, the dependence of the current gain factor $\beta(T)$ at a fixed value of a collector current is a decreasing function. This is a characteristic feature of bipolar power transistors made with silicon carbide technology [8,25–27]. The results of simulations using the proposed β model has a good agreement with the results of the measurements of the $\beta(i_C)$ characteristics in all ranges of temperature, which indicates that the transistor's current gain properties deteriorate with the temperature increase.

Figure 7 presents the isothermal and non-isothermal dependence of the current gain factor β of the considered transistor on the ambient temperature at a fixed value of the collector-emitter voltage and the control current equal to 3 V and 50 mA, respectively. The points represent the results of measurements and lines represent the results of calculations using the proposed ETM.

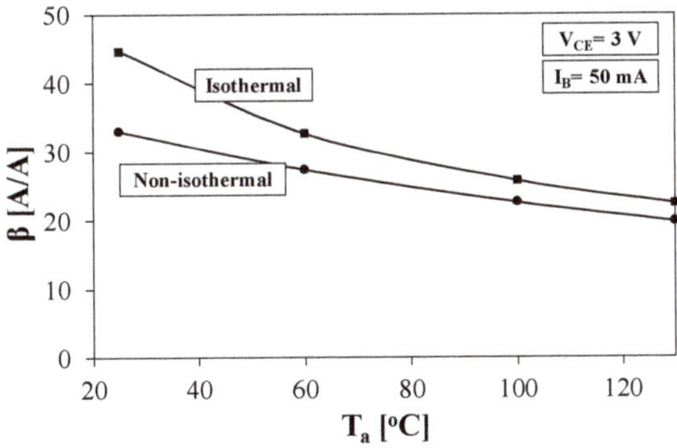

Figure 7. Isothermal and non-isothermal dependence of the current gain factor.

As seen, the non-isothermal characteristic of SiC BJT lies under the isothermal counterpart. Moreover, at room temperature the differences between isothermal and non-isothermal measurements reach over 25%. This phenomenon is due to the fact that when the device lattice temperature and phonon scattering in SiC BJTs increase, they cause both the carrier mobility and the current gain to drop [28].

4. Conclusions

The authors' measurements show that the ambient temperature rises strongly affect certain characteristics and parameters of the silicon carbide power bipolar transistor. Moreover, the self-heating phenomenon in SiC BJT can result in not only qualitative but also quantitative changing of the semiconductor device characteristics shape, resulting in the negative slope of collector current with increasing collector-emitter voltage occurrence and causes decreasing of the current gain factor β with power dissipation increase.

In the typical Gummel–Poon model implemented in SPICE, the self-heating phenomenon is not taken into account. In this paper, the authors proposed an improved SiC BJT model dedicated for SPICE, incorporating not only the self-heating phenomenon, but also modeling the quasi–saturation mode using the temperature-dependent intrinsic carrier concentration model. Apart from this, the current gain factor model is used for the first time for the bipolar transistor made of silicon carbide taking the temperature and collector current impact into account. The proposed electrothermal model of the BT1206-AC SiC BJT allows calculating the non-isothermal characteristics of the tested device. The simulated and measured characteristics of considered SiC power BJT were compared. The satisfactory agreement between both the calculated and measured characteristics indicates the correctness of the presented model. It is worth mentioning that the accuracy of the proposed model can be achieved even at high current density when the current gain factor β starts to decrease.

Author Contributions: Investigation, J.P., D.B. and J.Z.; methodology, J.P., D.B. and J.Z.; supervision, J.Z.; writing—original draft, J.P., D.B. and J.Z.; writing—review and editing, J.P., D.B. and J.Z. All authors have read and agreed to the published version of the manuscript.

Funding: This research received no external funding.

Conflicts of Interest: The authors declare no conflicts of interest.

References

1. Wang, J.; Liang, S.; Deng, L.; Yin, X.; Shen, Z.J. An improved SPICE model of SiC BJT incorporating surface recombination effect. *IEEE Trans. Power Electron.* **2019**, *34*, 6794–6802. [CrossRef]
2. Bargieł, K.; Bisewski, D.; Zarębski, J. Modelling of dynamic properties of silicon carbide junction field-effect transistors (JFETs). *Energies* **2020**, *13*, 187. [CrossRef]
3. Zarębski, J.; Dąbrowski, J.; Bisewski, D. Measurements of thermal parameters of silicon carbide semiconductor devices. *Przegląd Elektrotechniczny* **2011**, *87*, 29–32.
4. DiMarino, C.; Chen, Z.; Boroyevich, D.; Burgos, R.; Mattavelli, P. Characterization and comparison of 1.2 kV SiC power semiconductor devices. In Proceedings of the 15th European Conference on Power Electronics and Applications (EPE), Lille, France, 2–6 September 2013; pp. 1–10.
5. Catalog data of SiC BJT BT1206-AC. Available online: https://www.dacpol.eu/pl/elementy-polprzewodnikowe-z-weglika-krzemu/product/elementy-polprzewodnikowe-z-weglika-krzemu-1224 (accessed on 20 February 2020).
6. Singh, R.; Grummel, B.; Sundaresan, S. Short circuit robustness of 1200 V SiC switches. In Proceedings of the IEEE 3rd Workshop on Wide Bandgap Power Devices and Applications (WiPDA), Blacksburg, VA, USA, 2–4 November 2015; pp. 1–4.
7. Oueslati, M.; Garrab, H.; Jedidi, A.; Besbes, K. The advantage of Silicon Carbide material in designing of power Bipolar Junction Transistors. In Proceedings of the 12th International Multi-Conference on Systems, Signals & Devices, Mahdia, Tunisia, 16–19 March 2015; pp. 1–6.
8. Patrzyk, J.; Zarębski, J.; Bisewski, D. DC characteristics and parameters of silicon carbide high-voltage power BJTs. In Proceedings of the 39th International Microelectronics and Packaging (IMAPS) Conference, Gdańsk, Poland, 20–23 September 2015; pp. 1–8.
9. Santi, E.; Peng, K.; Mantooth, H.A.; Hudgins, J.L. Modeling of wide-bandgap power semiconductor devices—part II. *IEEE Trans. Electron. Devices* **2015**, *62*, 434–442. [CrossRef]
10. Górecki, K.; Detka, K. Influence of power losses in the inductor core on characteristics of selected DC-DC converters. *Energies* **2019**, *12*, 1991. [CrossRef]

11. Dąbrowski, J.; Krac, E.; Górecki, K. New model of solar cells for SPICE. In Proceedings of the 25th International Conference Mixed Design of Integrated Circuits and Systems MIXDES, Gdynia, Poland, 21–23 June 2018; pp. 338–342.
12. Janicki, M.; Torzewicz, T.; Ptak, P.; Raszkowski, T.; Samson, A.; Górecki, K. Parametric compact thermal models of power LEDs. *Energies* **2019**, *12*, 1724. [CrossRef]
13. Mantooth, H.A.; Peng, K.; Santi, E.; Hudgins, J.L. Modeling of wide bandgap power semiconductor devices-part I. *IEEE Trans. Electron. Devices* **2015**, *62*, 423–433. [CrossRef]
14. Tian, Y.; Hedayati, R.; Zetterling, C.-M. SiC BJT Compact DC model with continuous-temperature scalability from 300 to 773 K. *IEEE Trans. Electron. Devices* **2017**, *64*, 3588–3594. [CrossRef]
15. Balachandran, S.; Chow, T.P.; Agarwal, A.; Tipton, W.; Scozzie, S. Gummel-Poon model for 1.8 kV SiC high-voltage bipolar junction transistor. In Proceedings of the IEEE 35th Annual PESC, Aachen, Germany, 20–25 June 2004; pp. 2994–2998.
16. Johannesson, D.; Nawaz, M. Development of a simple analytical PSpice model for SiC-based BJT power modules. *IEEE Trans. Power Electron.* **2016**, *31*, 4517–4525. [CrossRef]
17. Liu, W. Electro-Thermal Simulations and Measurements of Silicon Carbide Power Transistors. Ph.D. Thesis, The Royal Institute of Technology, Stockholm, Sweden, 2004.
18. Amin, A.R.; Salehi, A.; Ghezelayagh, M.H. BJT circuits simulation including self-heating effect using FDTD method. In Proceedings of the Asia-Pacific International Symposium on Electromagnetic Compatibility, Beijing, China, 12–16 April 2010; pp. 924–927.
19. Cadence Design Systems Inc. *PSPICE A/D Reference Guide Version 10.2*; Cadence Design Systems Inc.: San Jose, CA, USA, 2004.
20. Zarębski, J.; Górecki, K. Parameters estimation of the d.c. electrothermal model of the bipolar transistor. *Int. J. Numer. Model. Electron. Netw. Devices Fields* **2002**, *15*, 181–194. [CrossRef]
21. Górecki, K.; Zarębski, J.; Górecki, P.; Ptak, P. Compact thermal models of semiconductor devices: A Review. *Int. J. Electron. Telecommun.* **2019**, *65*, 151–158.
22. Górecki, P.; Górecki, K. Modelling a switching process of IGBTs with influence of temperature taken into account. *Energies* **2019**, *12*, 1894. [CrossRef]
23. Bisewski, D. Parameters estimation of SPICE models for silicon carbide devices. In Proceedings of the 21st European Microelectronics and Packaging Conference (EMPC) and Exhibition, Warszawa, Poland, 10–13 September 2017; pp. 1–5.
24. Feng, Z.; Zhang, X.; Wang, J.; Yu, S. High-efficiency three-level ANPC inverter based on hybrid SiC and Si devices. *Energies* **2020**, *13*, 1159. [CrossRef]
25. Daranagama, T.; Pathirana, V.; Udrea, F.; McMahon, R. Novel 4H-SiC bipolar junction transistor (BJT) with improved current gain. In Proceedings of the 13th Brazilian Power Electronics Conference and 1st Southern Power Electronics Conference (COBEP/SPEC), Fortaleza, Brazil, 28 November–2 December 2015; pp. 1–6.
26. Elahipanah, H.; Kargarrazi, S.; Salemi, A.; Östling, M.; Zetterling, C.-M. 500 °C High current 4H-SiC lateral BJTs for high-temperature integrated circuits. *IEEE Electron. Device Lett.* **2017**, *38*, 1429–1432. [CrossRef]
27. Liang, S.; Wang, J.; Peng, Z.; Chen, G.; Yin, X.; Shen, Z.J.; Deng, L. A modified behavior spice model for SiC BJT. In Proceedings of the IEEE Applied Power Electronics Conference and Exposition (APEC), San Antonio, TX, USA, 4–8 March 2018; pp. 238–243.
28. Meng, F.; Ma, J.; He, J.; Li, W. Phonon-limited carrier mobility and temperature-dependent scattering mechanism of 3 C-SiC from first principles. *Phys. Rev. B* **2019**, *99*, 045201. [CrossRef]

Article

Analysis of Algorithm Efficiency for Heat Diffusion at Nanoscale Based on a MEMS Structure Investigation

Tomasz Raszkowski * and Mariusz Zubert

Department of Microelectronics and Computer Science, Lodz University of Technology, 90-924 Lodz, Poland; mariuszz@dmcs.p.lodz.pl
* Correspondence: traszk@dmcs.pl

Received: 25 March 2020; Accepted: 13 May 2020; Published: 15 May 2020

Abstract: This paper presents an analysis of the time complexity of algorithms prepared for solving heat transfer problems at nanoscale. The first algorithm uses the classic Dual-Phase-Lag model, whereas the second algorithm employs a reduced version of the model obtained using a Krylov subspace method. This manuscript includes a description of the finite difference method approximation prepared for analysis of the real microelectromechanical system (MEMS) structure manufactured by the Polish Institute of Electron Technology. In addition, an approximation scheme of the model, as well as the Krylov subspace-based model order reduction technique are also described. The paper considers simulation results obtained using both investigated versions of the Dual-Phase-Lag model. Moreover, the relative error generated by the reduced model, as well as the computational complexity of both algorithms, and a convergence of the proposed approach are analyzed. Finally, all analyses are discussed in detail.

Keywords: Dual-Phase-Lag heat transfer model; Krylov subspace-based model order reduction; algorithm efficiency analysis; relative error analysis; algorithm convergence analysis; computational complexity analysis; thermal simulation algorithm; finite difference method scheme; Grünwald–Letnikov fractional derivative

---

1. Classical and Modern Heat Transfer Description

Currently, deep development in technology has increased interest in heat transfer modeling. New thermal problems have been observed due to the continuous reduction of the size of electronic devices or miniaturization of integrated circuits. For instance, in such small electronic appliances, a significant rise of operational frequency and a rapid increase of generated internal heat density have been noticed which have a meaningful influence on temperature rise during the operation of the device.

It is worthwhile to highlight that it is extremely important to ensure the appropriate cooling conditions, however, this issue is very problematic in the case of nanosized electronic devices and can result in unstable and improper operation of the device. Moreover, most of the damages and malfunctions of the device are caused by unsuitable operation and thermal problems. Thus, thermal analysis is very important and is one of the most crucial steps in the design and planning of modern electronic appliances.

The heat transfer problems have been modeled using Fourier's theory [1]. This method is based on Fourier's law and the Fourier–Kirchhoff (FK) equation, and can be expressed as follows:

$$\begin{cases} \boldsymbol{q}(\boldsymbol{x},t) = -k \cdot \nabla T(\boldsymbol{x},t) \\ c_v \frac{\partial T(\boldsymbol{x},t)}{\partial t} + \nabla \circ \boldsymbol{q}(\boldsymbol{x},t) = q_V(\boldsymbol{x},t) \end{cases} \qquad \boldsymbol{x} \in R^n, n \in N, t \in R_+ \cup \{0\} \tag{1}$$

In the above system of equations, the heat flux density vector is represented by $\boldsymbol{q}$, thermal conductivity of analyzed material is expressed by k, and T describes the function of temperature rise. Moreover, the volumetric heat capacity and value of internally generated heat are represented by c_v and q_V, respectively.

The Fourier–Kirchhoff method has been successfully applied for two centuries. However, in modern nanosized structures, application of the FK method has some serious limitations [2–4]. First, the assumption of infinite speed propagation of the heat is postulated. Moreover, the instantaneous change of heat flux or temperature gradient should also be taken into account as a non-physical behavior of the structures. The phenomena mentioned have not been empirically proven in the case of nanosized electronic structures [5,6]. Thus, a new methodology, which significantly improves the FK method in the case of modern electronic structures, should be established.

In the mid-1990s, a new approach, called Dual-Phase-Lag (DPL), was introduced by Tzou as a more appropriate choice for modeling temperature changes in a nanosized electronic structure [7]. The DPL model based on the FK theory, however, as previously mentioned, includes improvements such as time lags [8]. These lags express the needed time change in the heat flux density, as well as the temperature gradient.

The lags mentioned above are represented by different values expressed by τ_q, which is related to heat flux time lag, and τ_T which represents the temperature time lag. Taking into consideration new time lags, the DPL model can be described by the following equations:

$$\begin{cases} \nabla \circ q(x,t) = -c_v \frac{\partial T(x,t)}{\partial t} + q_V(x,t) \\ k\tau_T \frac{\partial \nabla T(x,t)}{\partial t} + \tau_q \frac{\partial q(x,t)}{\partial t} = -k \cdot \nabla T(x,t) - q(x,t) \end{cases} \quad x \in R^n, n \in N, t \in R_+ \cup \{0\} \tag{2}$$

The DPL model can be successfully applied instead of the classical FK model due to the fact that it is appropriate for parabolic partial differential equations, as well as for hyperbolic equations (see also [9,10]).

However, some disadvantages have also appeared. The DPL model is classified as a model with greater complexity than the FK model. Thus, to carry out the simulation, a longer computation time is needed, especially for complex electronic structures.

Considering the above limitation of the DPL model, the main aim of this paper is to implement a DPL-based method which reduces the time of simulation and can be as accurate as the original DPL model. One approach which can be applied is the Krylov subspace-based model order reduction method [11]. This approach significantly reduces the number of equations in the system describing an analyzed heat transfer problem.

The Krylov subspace method is used for the order reduction of large-scale systems, especially in a mechanical domain (e.g., J. White and J. G. Korvink papers, as well as [12,13]). However, the research presented in this paper is focused on aspects other than those of previously published papers. First of all, this manuscript includes the first application of the Krylov subspace-based method for the DPL heat transfer equation. Moreover, in previous research, the spectrum of mechanical systems generally contains frequencies from 0 to hundreds of Hz (sometimes up to 1–5 kHz). The model is validated typically for longer times, for example, 10 ms–1 s. However, our research includes significantly greater frequency values resulting in meaningfully shorter times and smaller DPL time lags, even hundreds of femtoseconds or a few nanoseconds. Therefore, the range of applicability of the described model order reduction methodology, presented in the manuscript, is different than other previous applications.

The structure of the paper is as follows: First, a short description of the investigated real test nanosized structure, its finite difference method approximation, and structure discretization are presented; then, the approximation scheme of the DPL model for the considered MEMS structure is proposed; after that, the Krylov subspace-based model order reduction technique is demonstrated; and finally, the simulation results obtained using both reduced and non-reduced versions of the DPL model are compared, analyzed, and discussed.

2. Mathematical Preliminaries of the Proposed Methodology

2.1. Finite Difference Method Approximation

Thermal analysis was carried out for the real microelectromechanical system (MEMS) nanostructure, presented in Figure 1, which was manufactured at the Polish Institute of Electron Technology [14,15]. This test structure includes two parallel platinum resistors, each 10 μm long. One resistor is treated as a heater, whereas the second resistor plays the role of a temperature sensor. The cross-sectional area of each resistor is 100 nm wide and 100 nm high. The distance between these resistors is 100 nm. The resistors are placed on a 100 nm wide silicon dioxide layer and both are stacked on a 500 μm thick silicon layer. A detailed description of the investigated structure, as well as the measurement process are found in [14,15].

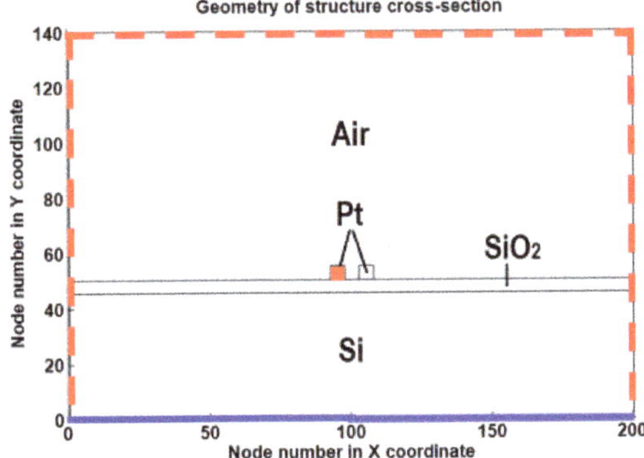

Figure 1. Cross-sectional area of the microelectromechanical system (MEMS) test structure where the Dirichlet boundary conditions are marked with a blue line, the Neumann boundary conditions are marked with a red dashed line, and the internal heat source is marked using red color.

In the investigated cross-sectional area of the MEMS test structure, the Dirichlet boundary conditions are at the bottom, while the Neumann boundary conditions are used on the left, right, and the top parts of the structure and their environment. The boundary conditions can be described by the following equations:

$$T_k(t) = 0, t \in R_+ \cup \{0\},$$
$$k \in \{1, 2, 3, \ldots, n_x\} \tag{3}$$

$$q_k(t) = 0, t \in R_+ \cup \{0\},$$
$$k \in \{n_x + 1, 2 \cdot n_x + 1, \ldots, (n_y - 1) \cdot n_x + 1\} \cup \{n_x, 2 \cdot n_x, \ldots, (n_y - 1) \cdot n_x\} \cup$$
$$\cup \{(n_y - 1) \cdot n_x + 1, (n_y - 1) \cdot n_x + 2, \ldots, n_y \cdot n_x\} \tag{4}$$

where n_x and n_y reflects the number of discretization nodes in both axes OX and OY, respectively.

Thermal simulations for the investigated MEMS structure were performed for its two-dimensional (2D) cross-sectional area, in the middle of the resistor. The thermal analysis of the structure was carried out using the DPL model. In order to make the analysis easier, the DPL model described by the system Equation (2) was equivalently transformed to the following 2D form [16]:

$$c_v \cdot \tau_q \cdot \frac{\partial^2 T(x,y,t)}{\partial t^2} + c_v \cdot \frac{\partial T(x,y,t)}{\partial t} - k \cdot \tau_T \cdot \frac{\partial \Delta T(x,y,t)}{\partial t} - k \cdot \Delta T(x, y, t) =$$
$$= q_V(x, y, t) + \tau_q \cdot \frac{\partial q_V(x,y,t)}{\partial t} \text{ for } x, y \in R, t \in R_+ \cup \{0\} \tag{5}$$

As presented in Appendix A, the term $\tau_q \frac{\partial q_v(t)}{\partial t}$ can be neglected with an error of about -1.91%, for the simulation time $t \geq 3\tau_T$ and $|err| \leq 0.58\%$ for $t \geq 10\tau_T$. The $q_v(t + \tau_q)$ can also be used instead of

$q_v(t) + \tau_q \frac{\partial q_v(t)}{\partial t}$ for $q_v(t) = c_1 \cdot H(t)$, where $H(t)$ is a Heaviside function, c_1 is a constant where $|c_1| < +\infty$. It is worthwhile highlighting that the presented values were obtained considering the temperature and heat flux time lags for platinum. The Laplacian ΔT was approximated using the finite difference method (FDM) approach for a rectangular mesh with a constant distance between nodes in both dimensions. The considered approximation is presented below:

$$\Delta T(x, y, t) \approx \frac{T(x+\Delta x,y,t)+T(x,y+\Delta x,t)-4 \cdot T(x,y,t)+T(x-\Delta x,y,t)+T(x,y-\Delta x,t)}{(\Delta x)^2},$$
$$\text{dla } x, y \in R, t \in R_+ \cup \{0\} \tag{6}$$

Then, the DPL Equation (5) becomes an ordinary differential equation (ODE) of a time variable. Such a constructed system of equations, including each investigated mesh node, was solved based on the backward differentiation formulas (BDF) approach [17–20], with initial conditions $T(x,y,t) = 0$ and $\partial T(x,y,t)/\partial t = 0$ for $t \leq 0$, which means zero initial conditions for Equation (16).

2.2. Description of Structure Discretization

The analyzed cross-sectional area of the MEMS structure was discretized using the mesh that can be described by the following equations [16]:

$$q_k(t) = q(x, y, t), x = i \cdot \Delta x, y = j \cdot \Delta y \tag{7}$$

$$T_k(t) = T(x, y, t), x = i \cdot \Delta x, y = j \cdot \Delta y \tag{8}$$

$$i \in \{1, 2, \dots, n_x\}, j \in \{1, 2, \dots, n_y\}, k \in \{1, 2, \dots, n_x \cdot n_y\}$$

In the Equations above, n_x and n_y reflect the number of discretization nodes in both axes. It means that the product $n_x \times n_y$ describes the number of nodes used in discretization mesh. Nodes are numbered from the left side to the right side. First, the first bottom row is considered. When all nodes in this row have already had their numbers, the procedure is repeated in the second row from the bottom side. This process is continued until all nodes in a top row are numbered. The graphical interpretation of the applied discretization mesh for the investigated cross-section of the MEMS structure, as well as nodes numbering approach, are demonstrated in Figure 2.

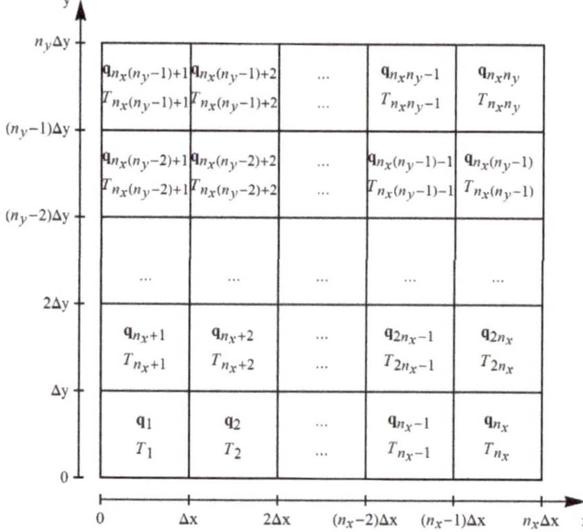

Figure 2. The graphical interpretation of the proposed discretization mesh nodes numbering [16,21].

2.3. DPL Model Approximation Scheme for the Investigated MEMS Structure

Considering the imposed FDM assumptions, the DPL model can be expressed in the following way [22]:

$$\begin{cases} \boldsymbol{M} \cdot \ddot{\boldsymbol{T}}(t) + \boldsymbol{D} \cdot \dot{\boldsymbol{T}}(t) + \boldsymbol{K} \cdot \boldsymbol{T}(t) = \boldsymbol{b} \cdot u(t) \\ \boldsymbol{y}(t) = \boldsymbol{c}^T \cdot \boldsymbol{T}(t) \end{cases} \quad t \in R_+ \cup \{0\} \tag{9}$$

where superscript T means a transposition. Moreover, T is the temperature function, while $\dot{T}$ and $\ddot{T}$ are its first and second time derivative, respectively. All of them are $n_x \cdot n_y \times 1$ vectors. Other vectors and matrices can be described as follows [23]:

$$\begin{aligned} &\boldsymbol{I}, \boldsymbol{M_{FDM}}, \boldsymbol{M}, \boldsymbol{D}, \boldsymbol{K}, \boldsymbol{c}^T \in R^{n_x \cdot n_y \times n_x \cdot n_y}, \\ &\boldsymbol{c_v}, \boldsymbol{k}, \boldsymbol{\tau_q}, \boldsymbol{\tau_T}, \boldsymbol{b}, \boldsymbol{y} \in R^{n_x \cdot n_y \times 1}, \\ &u \in R \end{aligned}$$

$$\boldsymbol{M} = diag\big(\boldsymbol{\tau_q}\big) \circ diag(\boldsymbol{c_v}) \circ \boldsymbol{I} \tag{10}$$

$$\boldsymbol{D} = diag(\boldsymbol{c_v}) \circ \boldsymbol{I} - \frac{1}{(\Delta x)^2} \cdot repmat(\boldsymbol{\tau_T}) \circ repmat(\boldsymbol{k}) \circ \boldsymbol{M_{FDM}} \tag{11}$$

$$\boldsymbol{K} = -\frac{1}{(\Delta x)^2} repmat(\boldsymbol{k}) \circ \boldsymbol{M_{FDM}} \tag{12}$$

$$\boldsymbol{b} = a \cdot \begin{bmatrix} 0 & \cdots & 0 & 1 & \cdots & 1 & 0 & \cdots & 0 \end{bmatrix}^T, a \in R_+ \tag{13}$$

$$\boldsymbol{c} = \boldsymbol{I} \tag{14}$$

where $\boldsymbol{I}$ is identity matrix, an operator $\circ$ means a Hadamard product, function diag($\cdot$) generates a diagonal matrix based on a given vector, and function repmat($\cdot$) replicates a considered vector and generates a matrix of desired dimensions. In addition, non-zero elements of vector $\boldsymbol{b}$ indicate nodes with internal heat generation. Moreover, $\boldsymbol{M_{FDM}}$ is a matrix including FDM coefficients wgugc us determined based on Equation (6) and the considered boundary conditions. It is also worthwhile to note that the $\boldsymbol{M_{FDM}}$ coefficients describing the air area between platinum resistors' surfaces, as well as the contact between them and the oxide, have been calculated considering a fractional order of the temperature function space derivative based on the Grünwald–Letnikov theory [22,24]. Thus, in this particular case, each one-dimensional part of Equation (6) is replaced by the following Equation [22]:

$$\begin{aligned} &\frac{1}{(\Delta s)^{\alpha_x}} \cdot \sum_{k=0}^{2} (-1)^k \frac{\Gamma(\alpha_x+1)}{\Gamma(k+1) \cdot \Gamma(\alpha_x-k+1)} T\Big(x - k \cdot \Delta x + \tfrac{\alpha_x \cdot \Delta x}{2}, t\Big) \cdot 1 = \\ &= \frac{\big(\tfrac{\alpha_x}{2}-1\big) \cdot T(x+2 \cdot \Delta x, y, t) + \big(2-\tfrac{\alpha_x}{2}-\alpha_x \cdot \big(\tfrac{\alpha_x}{2}-1\big)\big) \cdot T(x+\Delta x, y, t)}{(\Delta x)^{\alpha_x}} + \\ &+ \frac{\big(\tfrac{\alpha_x \cdot (\alpha_x-1)}{2} \cdot \big(\tfrac{\alpha_x}{2}-1\big) - \alpha_x \cdot \big(2-\tfrac{\alpha_x}{2}\big)\big) \cdot T(x,y,t) + \big(\big(2-\tfrac{\alpha_x}{2}\big) \cdot \tfrac{\alpha_x \cdot (\alpha_x-1)}{2}\big) \cdot T(x-\Delta x, y, t)}{(\Delta x)^{\alpha_x}} + \\ &+ \frac{\big(\tfrac{\alpha_x}{2}-1\big) \cdot T(x, y+2 \cdot \Delta x, t) + \big(2-\tfrac{\alpha_x}{2}-\alpha_x \cdot \big(\tfrac{\alpha_x}{2}-1\big)\big) \cdot T(x, y+\Delta x, t)}{(\Delta x)^{\alpha_x}} + \\ &+ \frac{\big(\tfrac{\alpha_x \cdot (\alpha_x-1)}{2} \cdot \big(\tfrac{\alpha_x}{2}-1\big) - \alpha_x \cdot \big(2-\tfrac{\alpha_x}{2}\big)\big) \cdot T(x,y,t) + \big(\big(2-\tfrac{\alpha_x}{2}\big) \cdot \tfrac{\alpha_x \cdot (\alpha_x-1)}{2}\big) \cdot T(x, y-\Delta x, t)}{(\Delta x)^{\alpha_x}} \end{aligned} \tag{15}$$

for $\alpha_x \in (2, 2.5)$, $\Delta x \to 0$

In order to make the analysis more clear, the system Equations (9) of the second-order equations has been transformed into the first-order Equations [21]:

$$\begin{cases} \boldsymbol{E} \cdot \dot{\overline{\boldsymbol{T}}}(t) = \boldsymbol{A} \cdot \overline{\boldsymbol{T}}(t) + \boldsymbol{B} \cdot u(t) \\ \boldsymbol{y}(t) = \boldsymbol{C}^T \cdot \overline{\boldsymbol{T}}(t) \end{cases} \quad t \in R_+ \cup \{0\} \tag{16}$$

where $\bar{T}$ and $\bar{\dot{T}}$ are $2 \cdot n_x \cdot n_y \times 1$ vectors. First $n_x \cdot n_y$ elements of $\bar{T}$ are the same, similar to the case of T vector, while its second part coincides with $\dot{T}$. The first half of $\bar{\dot{T}}$ includes elements of $\dot{T}$, whereas its second half states $\ddot{T}$. In addition, the remaining matrices and vectors visible in Equation (16) present as follows [21,23,25]:

$$E = \begin{bmatrix} I & \Theta \\ \Theta & M \end{bmatrix}, A = \begin{bmatrix} \Theta & I \\ -K & -D \end{bmatrix}, B = \begin{bmatrix} \Theta_1 \\ b \end{bmatrix}, C = \begin{bmatrix} c & \Theta \\ \Theta & c \end{bmatrix} \tag{17}$$

where Θ and Θ_1 are null matrices. The final solution is determined using the BDF method with a variable order between 1 and 5.

2.4. Krylov Subspace Method

As seen in Figure 2, the thermal characteristic of an electronic structure can refer to the construction of a system consisting of a high number of differential equations. Moreover, the larger the system the more accurate the results. However, preparation of a numerical solution of such a system can be very time-consuming and can require significant computational power. Thus, a solution for this problem is needed. One idea to deal with it is an order reduction of a constructed system of equations.

The order reduction process can be based on a moment matching method [26]. This method helps with moments calculation, being a negative coefficients of a system's transfer function F_T in a Taylor's series, around point 0 [26,27]. Its form can be as follows [25]:

$$F_T(l) = -\sum_{i=0}^{\infty} m_i \cdot l^i \tag{18}$$

where m_i means an ith moment. For example, for the system Equation (16), moments are determined according to the following Equation:

$$m_i = C^T \left(A^{-1} \cdot E \right)^i \cdot A^{-1} \cdot B, i \in N \cup \{0\} \tag{19}$$

The main idea of the model order reduction is related to finding a new system of equations, which is characterized by a significantly lower order than the original one. In this process, it is extremely important to consider the same transfer function F_T for both systems. It assures the existence of the same initial moments in the original and reduced systems of equations. However, due to the form of Equation (18), a direct numerical determination of such a proposed solution is impossible. Thus, to resolve this problem, the Krylov subspace method can be used.

In linear algebra, the Krylov space (or subspace) K of an order r for a certain $n \times n$ square matrix U and n-element vector s is a linear subspace of a space R^n that is generated by the following vectors: $s, U \cdot s, U^2 \cdot s, \ldots, U^{n-1} \cdot s$. Thus, the following Equation is true [11,28,29]:

$$K_r(U, s) = \text{span}\left\{ s, U \cdot s, U^2 \cdot s, \ldots, U^{n-1} \cdot s \right\} \tag{20}$$

In the case of the previously analyzed system Equation (16), the matrix U is the same as A, while vector s is similar to B.

A Krylov subspace-based method is a numerical method that can find eigenvalues of large sparse matrices or solves a system with a high number of linear equations based on multiplications of vectors and matrices and operates on the determined vectors without the necessity of making additional operations on many matrices at the same time. Thus, starting with the vectors, the following vectors are determined, respectively: $U \cdot s, U^2 \cdot s$, etc.

Taking into consideration that consecutively determined vectors become linearly dependent relatively quickly, Krylov subspace-based methods require an additional application of some orthogonalization schemes. One of the most common of them are algorithms created by Lanczos [30–34], and Arnoldi [35–41]. The second algorithm mentioned was used in considerations presented in this paper. It can generate a set of orthonormal vectors which are determined using the modified Gram–Schmidt process (MGS) [42,43]. These vectors form a base Equation (20) of a Krylov subspace. Moreover, each vector from this base, starting with b, states consecutive columns of so-called transfer matrices V and W, which are used to determine coefficients matrices for reduced systems of linear equations. The algorithm stops after generating the first zero vectors, i.e., if the aggregated sum of absolute values of all vector's coordinates does not exceed the tolerance value equal to 0.0001. The number of algorithm's iterations (or the number of non-zero vectors) is equal to the reduced model order r. It is worthwhile highlighting that these zero vectors are not included in constructed V and W matrices.

In the investigated case, the V and W matrices are identical. Moreover, the number of iterations of the considered algorithm determines an order of the Krylov subspace and, at the same time, an order of a newly generated, reduced system of equations. For example, the reduced version of the system of Equation (16) is as follows [25]:

$$\begin{cases} E_r \cdot \dot{\overline{T}}_r(t) = A_r \cdot \overline{T}_r(t) + B_r \cdot u(t) \\ \qquad y(t) = C_r \cdot \overline{T}_r(t) \end{cases} \quad t \in R_+ \cup \{0\} \tag{21}$$

where the system Equation (21) is solved using backward differentiation formulas (BDFs) of variable order between 1 and 5. Previous research has shown that it is one of the most effective methods for solving these types of equation systems 3.

3. Thermal Simulation Results

3.1. Pre-Simulation Assumptions

Thermal simulation of the investigated MEMS structure was carried out using Matlab environment and proposed approximation scheme for the DPL model. The simulation results were calculated using the following material thermal parameters [23,44] included in Table 1:

Table 1. Considered material parameters' values [23,44].

Material Name	$k[\frac{W}{m \cdot K}]$	$\rho[\frac{kg}{m^3}]$	$c_p[\frac{J}{kg \cdot K}]$
Silicon (Si)	148	2 330	712
Silicon dioxide (SiO$_2$)	1.38	2220	745
Platinum (Pt)	71.6	21 450	133
Platinum (Pt)	71.6	21 450	133
Air	0.0263	1.1614	1.007

Moreover, based on [23,45], the DPL model parameters were estimated. The platinum resistors are characterized by the heat flux time lag approximately equal to 550 ps and the temperature time lag established at 15.6 ns. For other materials, these values are set to 18 ns and 480 ns, respectively. In order to make an analysis easier, all results were normalized according to the following Equation:

$$T_k^{norm}(t) = \frac{T_k(t)}{\max_{t,k}\{T_k(t)\}} k \in \{1, 2, \ldots, n_x \cdot n_y\}, t \in R_+ \cup \{0\} \tag{22}$$

3.2. Simulation Results

The thermal analyses of the investigated MEMS structure were divided into different areas. The first one, presented in this subsection, is related to the comparison between the results obtained using the reduced and non-reduced DPL model. In this area, the dynamic of average temperature rises in the platinum heater and the temperature sensor were investigated. Moreover, the temperature distribution in an entire cross-sectional area was considered for selected time points. It is also worthwhile highlighting that all results included in this subsection were received using a 10 nm distance between mesh nodes. Additionally, a comparison with FK and the measurement results was also carried out.

A comparison of the normalized average temperature rises over the time in the platinum heater and temperature sensor obtained using the reduced and non-reduced DPL model, the FK model, as well as real measurements is shown in Figure 3. It can be seen that there are almost no differences between the results plotted based on the reduced and non-reduced DPL models. The black solid line, which indicates the non-reduced DPL model results for the heater, coincide almost exactly with the red dashed line, which shows the outputs of the reduced DPL approach. A similar situation is also visible for the case of the temperature sensor. The black dotted line shows the results yielded using the non-reduced DPL model, and it coincides with the dashed blue curve, which shows outputs produced by the reduced DPL approach.

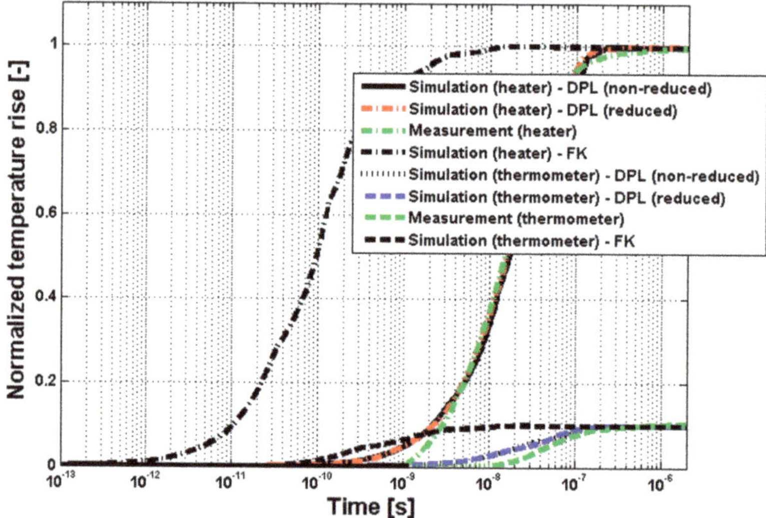

Figure 3. Comparison of normalized average temperature rises over the time in the platinum heater and the temperature sensor obtained using the reduced and non-reduced Dual-Phase-Lag (DPL) model, Fourier–Kirchhoff (FK) model, and measurements.

For comparison purposes, the measurement results which are indicated by the green lines were also plotted. It can be seen that the volatility over the time is very similar to the simulation results obtained using the DPL model, which confirms the correctness of the proposed approach.

The FK model produces significantly different outputs. The dashed and dotted black curve, which shows the temperature rise in the heater, as well as the dashed black line, which indicates the temperature changes over time in the temperature sensor, does not coincide with the DPL model or the measurement results. Thus, the FK model should not be used for temperature distribution determination at nanoscale.

In addition, an analysis of a temperature distribution inside an entire considered cross-sectional area of the MEMS structure was carried out. The results' comparison, for selected time points, is demonstrated in Figures 4–6.

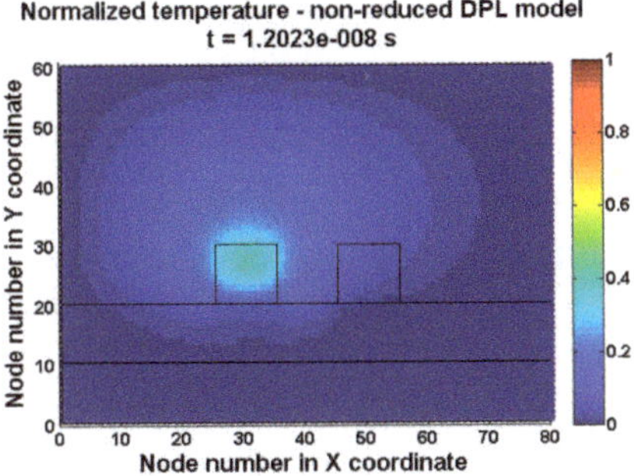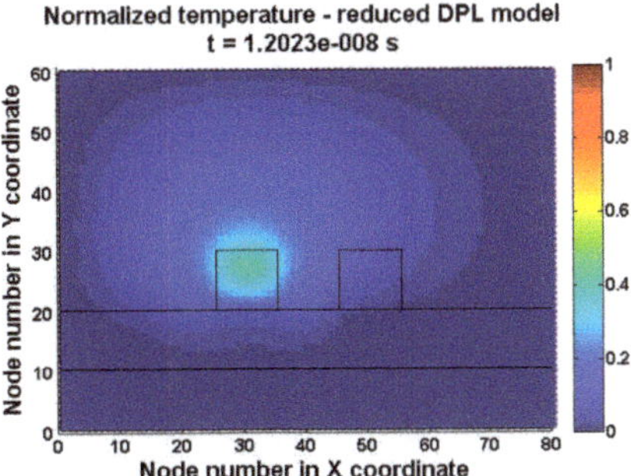

Figure 4. Comparison of normalized temperature rise in a cross-sectional area of the investigated MEMS structure obtained using reduced and non-reduced DPL model for t = 12.023 ns.

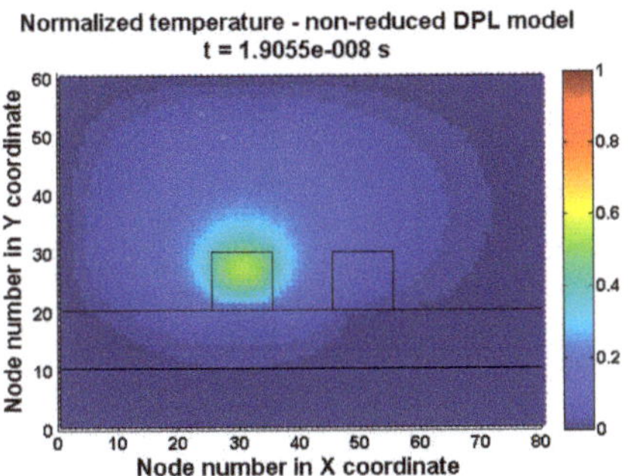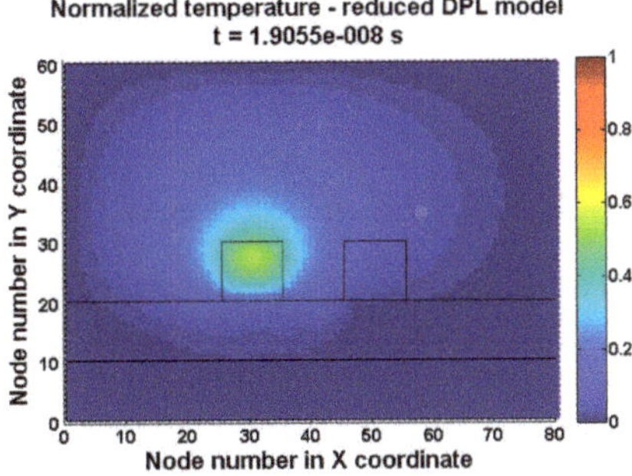

Figure 5. Comparison of normalized temperature rise in a cross-sectional area of the investigated MEMS structure obtained using reduced and non-reduced DPL model for t = 19.055 ns.

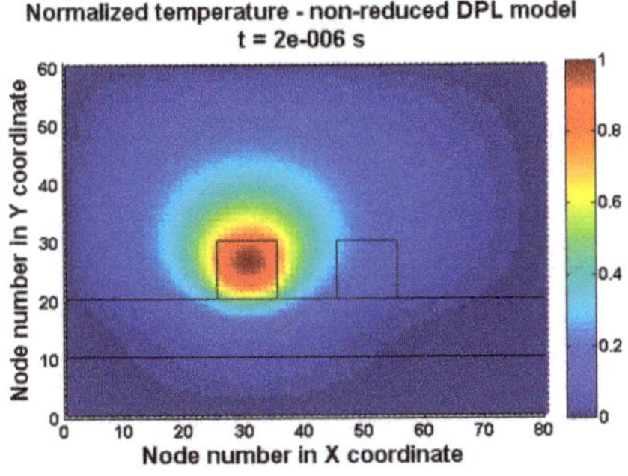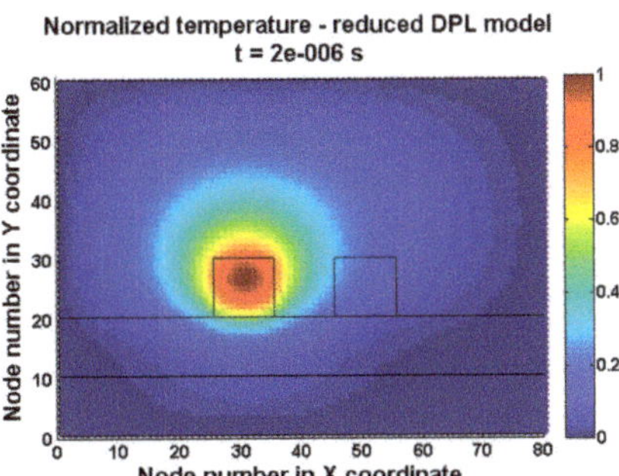

Figure 6. Comparison of normalized temperature rise in a cross-sectional area of the investigated MEMS structure obtained using reduced and non-reduced DPL model for t = 2 µs (steady state).

For each investigated time point, i.e., at different stages of temperature rise, it can be seen that the differences between the results obtained using the reduced and non-reduced DPL model are almost unnoticeable, which also suggests a very good level of coincidence between both investigated DPL versions.

3.3. Relative Error Analysis

In order to assess the quality of results obtained using the reduced DPL model, an error analysis was carried out. Figure 7 shows a comparison of relative error values calculated between the reduced and non-reduced DPL model outputs for three selected points of time, similar to those investigated in the previous subsection. For all the mentioned time points, the relative error values in relation to each discretization mesh nodes were analyzed.

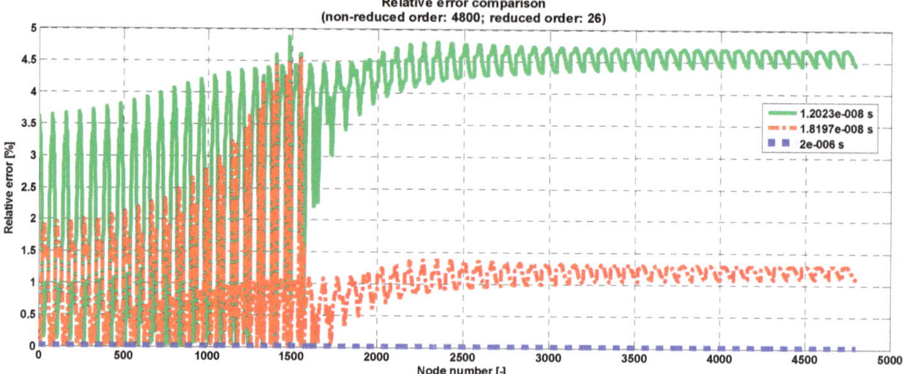

Figure 7. Comparison of relative error values in relation to a number of discretization mesh node for selected time points (t = 12.023 ns, t = 18.197 ns, and t = 2 μs).

The maximum relative error value did not exceed a level of 5%, which confirms good accuracy of the reduced DPL model results. Moreover, the relative error decreased with an increase in the value of the investigated time points. For example, for higher observed temperature rise values, the relative error is at a level of 1%–1.35%. Furthermore, for steady state, a maximum relative error value is approximately equal to 0.01%. Such a small error value can be neglected. The decrease of the relative error value over time also suggests a convergence of the proposed approach.

3.4. Computational Complexity Analysis

Finally, a computational complexity of the non-reduced and reduced DPL models was investigated. First, an analysis of the time consumed during the solution calculation was carried out and is shown in Figure 8. It can be seen that the smaller the distance between mesh nodes the longer the time needed to obtain a temperature distribution in the investigated cross-sectional area of the MEMS structure. This is caused by a greater number of nodes in the smaller considered node distances.

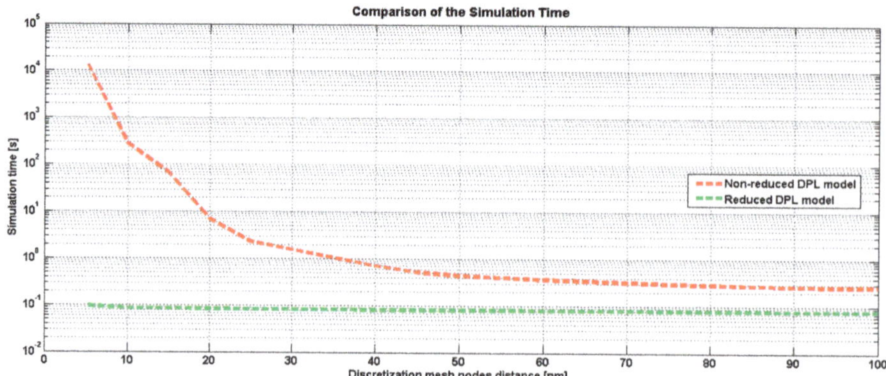

Figure 8. Comparison of simulation time consumed during solution calculation in relation to discretization mesh nodes distance based on non-reduced and reduced DPL models.

The simulation time investigation in relation to the number of nodes in the analyzed cross-section is shown in Figure 9. The significant difference in simulation time needed for results preparation is visible. Considering the non-reduced DPL model, on the one hand, the simulation time can be estimated by a power function according to Equation (23). On the other hand, the time consumed using the reduced DPL model can be approximated based on a linear function presented in Equation (24). Moreover, statistics describing Equations for simulation time estimations are shown in Table 2.

$$t^{DPL}_{non-reduced}(n) = 1.96 \cdot 10^{-7} \cdot n^{2.544} \tag{23}$$

$$t^{DPL}_{reduced}(n) = 2.546 \cdot 10^{-5} \cdot n - 0.03782 \tag{24}$$

where n reflects the number of nodes in the considered MEMS structure cross-section.

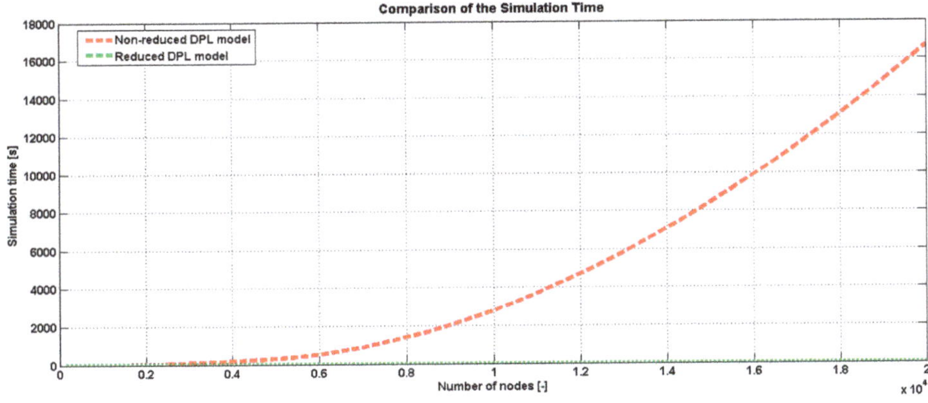

Figure 9. Comparison of the simulation time consumed during the solution calculation in relation to the number of nodes in the analyzed cross-section of the MEMS structure based on the non-reduced and reduced DPL models.

Table 2. Simulation time approximation.

Metrics	Adjusted R^2	R^2	RMSE
$t^{DPL}_{non-reduced}$	0.9988	0.9988	174.3
$t^{DPL}_{reduced}$	0.8299	0.829	0.06634

The values of metrics presented in Table 2 confirm the high quality of the prepared Equations for simulation time estimation. Moreover, based on Equations (23) and (24) and the simulation time analysis, it can be stated that the time complexity of an algorithm using the non-reduced DPL model is O($n^{2.544}$), whereas the reduced DPL is characterized by a time complexity O(n). It is also worthwhile highlighting that the DPL system Equation (21) is linear (regarding the time variable), and therefore the Krylov subspace is calculated only once. Thus, it is not updated in each time step. Of course, the computational complexity of the method includes considerations regarding the Krylov subspace generation.

4. Conclusions

This paper includes an analysis of the quality and time complexity of two algorithms dedicated to solving heat transfer problems at nanoscale. The first one uses the modern DPL model which is a significant improvement as compared with one of the most common approaches based on the FK model. The second one also employs the DPL model, however in its reduced version. The DPL model order reduction is prepared based on the Krylov subspace method.

The analyses have shown that both the reduced and non-reduced DPL models produce high quality results that coincide with real measurements of the test structure. Moreover, the results obtained using the reduced DPL model are very similar to the results yielded based on the non-reduced DPL approach. It is also worthwhile highlighting that the relative error of approximation of temperature distribution determination inside the test structure, which was obtained using the reduced DPL model, was at a very low level. Furthermore, this error decreased over the time, which suggested a convergence of the proposed approach.

In addition, the reduced DPL model prepares a solution for a heat transfer problem significantly faster than the classic version of this heat transfer model. The time complexity of the non-reduced approach is $O(n^{2.544})$, whereas in the case of the reduced model, the complexity is $O(n)$ only. Considering all the mentioned facts, it can be stated that the proposed approach obtained the high quality solution of the temperature distribution at nanoscale in a significantly shorter time than the classic approach, which is especially important to the future of designing and investigating advanced nanosized electronic structures.

Author Contributions: The algorithm for the FDM approximation scheme of reduced and non-reduced DPL model for 2D cross-section of analyzed test structure, investigation of the fractional Grünwald-Letnikov temperature derivative, application of Krylov subspace-based model order reduction technique in the case of DPL model, numerical simulations and evaluation of their results, relative error as well as computational complexity analysis of prepared algorithms and preparation of this manuscript have been carried out by T.R., M.Z. investigated algorithm convergence, the analysis of simplification error for internal heat generation source, supervised the research, and made corrections to the manuscript. All authors have read and agreed to the published version of the manuscript.

Funding: The research presented in this paper was carried out within the Polish National Science Centre project OPUS No. 2016/21/B/ST7/02247.

Acknowledgments: The authors would like to express their special thanks to M. Janicki and J. Topilko for sharing papers 14.

Conflicts of Interest: The authors declare no conflicts of interest.

Appendix A. Analysis of Internal Heat Generation Source in DPL Model

The DPL equation contains term $q_v(t) + \tau_q \frac{\partial q_v(t)}{\partial t}$, which is required for the accurate modeling of heat generation sources:

$$c_v \tau_q \frac{\partial^2 T}{\partial t^2} + c_v \frac{\partial T}{\partial t} = k\Delta T + k\tau_T \frac{\partial T}{\partial t} + \left(q_v + \tau_q \frac{\partial q_v}{\partial t}\right) \tag{A1}$$

Hence, the Taylor series expansions mapping of the heat generation in the DPL Equation (A1) into a time-shifted function 23 is considered as presented in Equation (A3). The higher order terms of the Taylor series expansions are neglected.

$$f(x,t) + \tau \frac{\partial f(x, t+\tau)}{\partial t} \rightarrow f(x, t+\tau) \tag{A2}$$

$$q_v(t) + \tau_q \frac{\partial q_v(t)}{\partial t} \approx q_v(t + \tau_q) \tag{A3}$$

After that, the accurate heat generation model $q_v(t) + \tau_q \frac{\partial q_v(t)}{\partial t}$ can be interpreted as a time-shifted function $q_v(t + \tau_q)$. It is sufficient to use a time-shifted delayed heat generation function $q_v(t + \tau_q)$ instead of term $q_v(t) + \tau_q \frac{\partial q_v(t)}{\partial t}$ in the case of absence of electro-thermal couplings ($\ldots \rightarrow q_v \rightarrow T \rightarrow q_v$ $\ldots$), as well as known excitation function $q_v(t)$, with the simulation time domain correction.

The advanced analysis based on the generalized functions theory 46 allows the estimation of relative error introduced by the simplified expression $q_v(t)$ instead of the $q_v(t) + \tau_q \frac{\partial q_v(t)}{\partial t}$ expression. Let us consider a more accurate model of internal heat source with the Heaviside excitation function $q_v(t) = H(t)$:

$$H(t) = \begin{cases} 1 & \text{for} \quad t > 0 \\ 0 & \text{for} \quad t < 0 \end{cases}, \text{ and } H'(t) = \delta(t) \tag{A4}$$

where $\delta(t)$ is the Dirac measure [46].

Suppose also $\varphi(t)$ is a border measured test function with compact support [46], then detailed probed calculated for internal heat source generalized function is described by the following Equation:

$$A = \left\langle q_v(t) + \tau_q \frac{\partial q_v(t)}{\partial t}, \varphi \right\rangle = \langle q_v, \varphi \rangle + \tau_q \left\langle \frac{\partial q_v}{\partial t}, \varphi \right\rangle = \langle q_v, \varphi \rangle - \tau_q \left\langle q_v, \frac{\partial \varphi}{\partial t} \right\rangle =$$
$$= \langle q_v, \varphi \rangle + \left\langle q_v, -\tau_q \frac{\partial \varphi}{\partial t} \right\rangle = \left\langle q_v, \varphi - \tau_q \frac{\partial \varphi}{\partial t} \right\rangle \tag{A5}$$

where

$$\langle f, \varphi \rangle = \int_{-\infty}^{+\infty} f(t)\,\varphi(t)dt \tag{A6}$$

Let us apply the same procedure for a simplified internal heat generation model for the DPL equation:

$$B = \langle q_v(t), \varphi \rangle \tag{A7}$$

Then, the relative error can be estimated by the following Equation:

$$err = \frac{B - A}{A} = \frac{-\tau_q \varphi(0)}{\int_0^{+\infty} \varphi(t)dt + \tau_q \varphi(0)} = \frac{-1}{\frac{1}{\tau_q} \int_0^{+\infty} \frac{\varphi(t)}{\varphi(0)} dt + 1} \approx \frac{-1}{\frac{1}{\tau_q} \int_0^{b} \frac{\varphi(t)}{\varphi(0)} dt + 1} \tag{A8}$$

where b is determined by the support of $\varphi(t)$. One of the most popular test function Equation (A9) has been used for the error evaluation *err*

$$\varphi(t) = \begin{cases} \exp\left(\frac{-a^2}{a^2 - t^2}\right) & \text{for} \quad |t| < a \\ 0 & \text{for} \quad |t| \geq a \end{cases} \tag{A9}$$

where a is an arbitrarily selected constant. The errors calculated for $a = \tau_q$ and several simulation times $b \in \{\tau_q, \tau_T, 3\tau_T, 10\tau_T\}$ are presented in Table A1. It can be seen that the simplified heat transfer model $q_v(t) = H(t)$ can be used, with relative error about -1.91%, for $a = \tau_q$, and simulation time about $t \geq 3\tau_T$, $b = 3\tau_T$. The relative error err is neglected for $t \geq 10\tau_T$ and $b = 10\tau_T$.

Table A1. Value of relative error calculated for different time values a, b, and the test function Equation (A9).

	$a = \tau_q, b = \tau_q$	$a = \tau_q, b = \tau_T$	$a = \tau_q, b = 3\tau_T$	$a = \tau_q, b = 10\tau_T$
	a) Derived from DPL model $B = \left\langle q_v(t) + \tau_q \frac{\partial q_v(t)}{\partial t}, \varphi \right\rangle \Rightarrow A = B$			
err =	0%	0%	0%	0%
	b) Approximation $B = \langle q_v(t), \varphi \rangle$			
err =	-62.37%	-5.52%	-1.91%	-0.58%
c) Approximation $B = \left\langle q_v(t + \tau_q), \varphi \right\rangle$ – It is the approximation of DPL equation, but it can be interpreted as the model in the real world.				
err =	-100%	-8.87%	-3.07%	-0.933%

It is worth emphasizing that calculations presented in the above table have been carried out for temperature and heat flux time lags for platinum.

References

1. Fourier, J.-B.J. *Théorie Analytique de la Chaleur*; Firmin Didot: Paris, France, 1822.
2. Raszkowski, T.; Samson, A. The Numerical Approaches to Heat Transfer Problem in Modern Electronic Structures. *Comput. Sci.* **2017**, *18*, 71–93. [CrossRef]

3. Raszkowski, T.; Zubert, M.; Janicki, M.; Napieralski, A. Numerical solution of 1-D DPL heat transfer equation. In Proceedings of the 22nd International Conference Mixed Design of Integrated Circuits and Systems (MIXDES), Torun, Poland, 25–27 June 2015; pp. 436–439.
4. Zubert, M.; Janicki, M.; Raszkowski, T.; Samson, A.; Nowak, P.S.; Pomorski, K. The Heat Transport in Nanoelectronic Devices and PDEs Translation into Hardware Description Languages. *Bulletin de la Société des Sciences et des Lettres de Łódź Série Recherches sur les Déformations* **2014**, *64*, 69–80.
5. Nabovati, A.; Sellan, D.P.; Amon, C.H. On the lattice Boltzmann method for phonon transport. *J. Comput. Phys.* **2011**, *230*, 5864–5876. [CrossRef]
6. Zubert, M.; Raszkowski, T.; Samson, A.; Janicki, M.; Napieralski, A. The distributed thermal model of fin field effect transistor. *Microelectron. Reliab.* **2016**, *67*, 9–14. [CrossRef]
7. Tzou, D.Y. A Unified Field Approach for Heat Conduction from Macro- to Micro-Scales. *J. Heat Transf.* **1995**, *117*, 8–16. [CrossRef]
8. Hays-Stang, K.; Haji-Sheikh, A. A unified solution for heat conduction in thin films. *Int. J. Heat Mass Transf.* **1999**, *42*, 455–465. [CrossRef]
9. Zhang, M.K.; Cao, B.Y.; Guo, Y.C. Numerical studies on dispersion of thermal waves. *Int. J. Heat Mass Transf.* **2013**, *67*, 1072–1082. [CrossRef]
10. Zhang, M.K.; Cao, B.Y.; Guo, Y.C. Numerical studies on damping of thermal waves. *Int. J. Therm. Sci.* **2014**, *84*, 9–20. [CrossRef]
11. Krylov, A.N. On the numerical solution of the equation by which in technical questions frequencies of small oscillations of material systems are determined. *Otdel. mat. i estest. nauk* **1931**, *7*, 491–539.
12. Lienemann, J.; Billger, D.; Rudnyi, E.B.; Greiner, A.; Korvink, J.G. MEMS Compact Modeling Meets Model Order Reduction: Examples of the Application of Arnoldi Methods to Microsystem Devices. In Proceedings of the 2004 Nanotechnology Conference and Trade Show, Nanotech 2004, Boston, MA, USA, 7–11 March 2004.
13. Benner, P.; Freund, R.W.; Sorensen, D.C. Special issue on Order reduction of large-scalesystems. *Linear Algebra Appl.* **2006**, *415*, 231–234. [CrossRef]
14. Sobczak, A.; Topilko, J.; Zajac, P.; Pietrzak, P.; Janicki, M. Compact Thermal Modelling of Nanostructures Containing Thin Film Platinum Resistors. In Proceedings of the 21st International Conference on Thermal, Mechanical and Multi-Physics Simulation and Experiments in Microelectronics and Microsystems (EuroSimE), Kraków, Poland, 26–29 April 2020.
15. Janicki, M.; Topilko, J.; Sobczak, A.; Zajac, P.; Pietrzak, P.; Napieralski, A. Measurement and Simulation of Test Structures Dedicated to the Investigation of Heat Diffusion at Nanoscale. In Proceedings of the 20th International Conference on Thermal, Mechanical and Multi-Physics Simulation and Experiments in Microelectronics and Microsystems (EuroSimE), Hannover, Germany, 24–27 March 2019. [CrossRef]
16. Raszkowski, T.; Samson, A.; Zubert, M. Investigation of Heat Distribution using Non-integer Order Time Derivative. *Bulletin de la Société des Sciences et des Lettres de Łódź Série Recherches sur les Déformations* **2018**, *68*, 79–92.
17. Curtiss, C.F.; Hirschfelder, J.O. Integration of stiff equations. *Proc. Natl. Acad. Sci. USA* **1952**, *38*, 235–243. [CrossRef] [PubMed]
18. Ascher, U.M.; Petzold, L.R. *Computer Methods for Ordinary Differential-Algebraic Equations*; SIAM: Philadelphia, PA, USA, 1998.
19. Süli, E.; Mayers, D. *An Introduction to Numerical Analysis*; Cambridge University Press: Cambridge, UK, 2003.
20. Auzinger, W.; Herfort, W.N. A uniform quantitative stiff stability estimate for BDF schemes. *Opusc. Math.* **2006**, *26*, 203–227.
21. Raszkowski, T.; Samson, A.; Zubert, M. Dual-Phase-Lag Model Order Reduction Using Krylov Subspace Method for 2-Dimensional Structures. *Bulletin de la Société des Sciences et des Lettres de Łódź Série Recherches sur les Déformations* **2018**, *68*, 55–68.
22. Raszkowski, T. Numerical Modelling of Thermal Phenomena in Nanometric Semiconductor Structures. Ph.D. Thesis, Lodz University of Technology, Lodz, Poland, 2019. (In Polish)
23. Raszkowski, T.; Zubert, M. Investigation of Heat Diffusion at Nanoscale based on Thermal Analysis of Real Test Structure. *Energies* **2020**, *13*, 2379. [CrossRef]
24. Raszkowski, T.; Samson, A.; Zubert, M. Temperature Distribution Changes Analysis based on Grünwald-Letnikov Space Derivative. *Bulletin de la Société des Sciences et des Lettres de Łódź Série Recherches sur les Déformations* **2018**, *68*, 141–152.

25. Raszkowski, T.; Samson, A.; Zubert, M.; Janicki MNapieralski, A. The numerical analysis of heat transfer at nanoscale using full and reduced DPL models. In Proceedings of the International Conference on Thermal, Mechanical and Multi-Physics Simulation and Experiments in Microelectronics and Microsystems (EuroSimE), Dresden, Germany, 3–5 April 2017.

26. Salimbahrami, B.; Lohmann, B. Order reduction of large scale second-order system using Krylov subspace methods. *Linear Algebra Appl.* **2006**, *415*, 385–405. [CrossRef]

27. Villemagne, C.D.; Skelton, R.E. Model reduction using a projection formulation. *Int. J. Control* **1987**, *46*, 2141–2169. [CrossRef]

28. Nevanlinna, O. *Convergence of Iterations for Linear Equations, Lectures in Mathematics ETH Zürich*; Birkhäuser Verlag: Basel, Switzerland, 1993.

29. Saad, Y. *Iterative Methods for Sparse Linear Systems*, 2nd ed.; Society for Industrial and Applied Mathematics: Philadelphia, PA, USA, 2003.

30. Lanczos, C. An iteration method for the solution of the eigenvalue problem of linear differential and integral operators. *J. Res. Natl. Bur. Stand.* **1950**, *45*, 255–282. [CrossRef]

31. Ojalvo, I.U.; Newman, M. Vibration modes of large structures by an automatic matrix-reduction method. *AIAA J.* **1970**, *8*, 1234–1239. [CrossRef]

32. Paige, C.C. Computational variants of the Lanczos method for the eigenproblem. *IMA J. Appl. Math.* **1972**, *10*, 373–381. [CrossRef]

33. Paige, C.C. The Computation of Eigenvalues and Eigenvectors of Very Large Sparse Matrices. Ph.D. Thesis, University of London, London, UK, 1971.

34. Ojalvo, I.U. Origins and advantages of Lanczos vectors for large dynamic systems. In Proceedings of the 6th Modal Analysis Conference (IMAC), Kissimmee, FL, USA, 1–4 February 1988; pp. 489–494.

35. Arnoldi, W.E. The principle of minimized iterations in solution of the matrix eigenvalue problem. *Q. Appl. Math.* **1951**, *9*, 17–29. [CrossRef]

36. Eid, R.; Salimbahrami, B.; Lohmann, B. Krylov-based order reduction using Laguerre series expansion. *Math. Comput. Model. Dyn. Syst.* **2008**, *14*, 435–449. [CrossRef]

37. Lehoucq, R.B.; Sorensen, D.C. *Deflation Techniques for an Implicitly Restarted Arnoldi Iteration*; SIAM: Philadelphia, PA, USA, 1996.

38. Lehoucq, R.B.; Sorensen, D.C.; Yang, C. *ARPACK Users Guide: Solution of Large-Scale Eigenvalue Problems with Implicitly Restarted Arnoldi Methods*; SIAM: Philadelphia, PA, USA, 1998.

39. Kokiopoulou, E.; Bekas, C.; Gallopoulos, E. *An Implicitly Restarted Lanczos Bidiagonalization Method for Computing Smallest Singular Triplets*; SIAM: Philadelphia, PA, USA, 2003.

40. Calvetti, D.; Reichel, L.; Sorensen, D.C. An Implicitly Restarted Lanczos Method for Large Symmetric Eigenvalue Problems. *Electron. Trans. Numer. Anal.* **1994**, *2*, 21.

41. Jia, Z. The refined harmonic Arnoldi method and an implicitly restarted refined algorithm for computing interior eigenpairs of large matrices. *Appl. Numer. Math.* **2002**, *42*, 489–512. [CrossRef]

42. Cheney, W.; Kincaid, D. *Linear Algebra: Theory and Applications*, 2nd ed; Sudbury, Jones & Bartlett: Sudbury, MA, USA, 2012.

43. Pursell, L.; Trimble, S.Y. Gram-Schmidt Orthogonalization by Gauss Elimination. *Am. Math. Mon.* **1991**, *98*, 544–549. [CrossRef]

44. Incropera, F.P.; DeWitt, D.P.; Bergman, T.L.; Lavine, A.S. *Fundamentals of Heat and Mass Transfer*, 6th ed.; John Wiley & Sons Inc.: Hoboken, NJ, USA, 2007.

45. Zubert, M.; Raszkowski, T.; Samson, A.; Zając, P. Methodology of determining the applicability range of the DPL model to heat transfer in modern integrated circuits comprised of FinFETs. *Microelectron. Reliab.* **2018**, *91*, 139–153. [CrossRef]

46. Grubb, G. *Distributions and Operators*; Springer: New York, NY, USA, 2000.

Article

Investigation of Heat Diffusion at Nanoscale Based on Thermal Analysis of Real Test Structure

Tomasz Raszkowski * and Mariusz Zubert

Department of Microelectronics and Computer Science, Faculty of Electrical, Electronic,
Computer and Control Engineering, Lodz University of Technology, 90-924 Lodz, Poland;
mariuszz@dmcs.p.lodz.pl
* Correspondence: traszk@dmcs.pl

Received: 18 April 2020; Accepted: 5 May 2020; Published: 9 May 2020

Abstract: This paper presents an analysis related to thermal simulation of the test structure dedicated to heat-diffusion investigation at the nanoscale. The test structure consists of thin platinum resistors mounted on wafer made of silicon dioxide. A bottom part of the structure contains the silicon layer. Simulations were carried out based on the thermal simulator prepared by the authors. Simulation results were compared with real measurement outputs yielded for the mentioned test structure. The authors also propose the Grünwald–Letnikov fractional space-derivative Dual-Phase-Lag heat transfer model as a more accurate model than the classical Fourier–Kirchhoff (F–K) heat transfer model. The approximation schema of proposed model is also proposed. The accuracy and computational properties of both numerical algorithms are presented in detail.

Keywords: Dual-Phase-Lag heat transfer model; thermal simulation algorithm; thermal measurements; Finite Difference Method scheme; Grünwald–Letnikov fractional derivative

---

1. Introduction

1.1. State of the Art

Nowadays, a heat transfer problem in the case of modern electronic structure is one of the most important research areas in the high-tech industry. The main reason for this fact is a significant downsizing of integrated circuits that is implemented in all modern electronic devices. Moreover, it also connects with a meaningful increase of an operation frequency of the mentioned devices. Both of these facts cause a rapid growth of a heat density generated inside such small structures. Consequently, the increased internal heat generation results in a huge increase of a temperature in critical parts of a device during its operation. It should also be highlighted that assurance of a proper cooling condition in the case of nanosized electronic structures is a non-trivial issue. Thus, all of these factors may influence an unstable operation and shorten the life cycle of an entire structure. Therefore, a thermal analysis seems to be one of the most important steps in the process of designing and developing modern electronic structures.

For a long time, heat conduction problems have been solved based on Fourier's theory [1,2]. This approach uses the Fourier's law and resulting Fourier–Kirchhoff (F–K) model for solids. They can be described by the following system of equations:

$$
\begin{cases}
\mathbf{q}(\mathbf{x},t) = -k \cdot \nabla T(\mathbf{x},t) \\
c_v \frac{\partial T(\mathbf{x},t)}{\partial t} + \nabla \circ \mathbf{q}(\mathbf{x},t) = q_V(\mathbf{x},t)
\end{cases}
\qquad \mathbf{x} \in R^n,\ n \in N,\ t \in R_+ \cup \{0\}
\tag{1}
$$

where $\mathbf{q}$ is a heat flux density vector; k means a thermal conductivity of investigated material; T is a function regarding temperature rise above the ambient temperature; c_v means a volumetric heat capacity being a product of a specific heat of a material for a constant pressure (c_p) and its density (ρ); and q_V reflects the value of internally generated heat.

The classical F–K equation can also be directly derived from the classical thermodynamics for a positive-definite entropy-production rate, for which the thermodynamic state transition in a heat transfer process is extremely slow (quasi-stationary process). Therefore, the process time (t) should be longer than a system's relaxation times.

Despite the numerous advantages of the Fourier–Kirchhoff approach, the research has shown that the application of the mentioned model may not reflect real phenomena [3–5]. One of its biggest disadvantages is an assumption considering the infinite speed of heat propagation. Apart from that, an instantaneous change of a temperature gradient or a heat flux should also be emphasized as a non-physical behavior postulated by the investigated theory. None of them, especially in the case of electronic structure with physical dimensions in nanometer-scale, has been empirically confirmed [6,7]. Thus, a thermal model including improvements of the F–K approach should be used instead of the classical theory.

The thermodynamic state is nonequilibrium in a fast transient process where time is comparable with the system relaxation times (e.g., an Umklapp phonon–phonon scattering process relaxation time in semiconductors [8,9]). The extended irreversible thermodynamic (EIT) can be applied to describe this kind of heat transfer system, assuming the second order the Taylor series expansion of entropy in EIT [10,11]. As a consequence, in mid-1990s, the new Dual-Phase-Lag (DPL) model was introduced by Tzou [12,13].

This model is an advanced version of an approach based on the Fourier–Kirchhoff theory and includes so-called time lags describing the time needed to change the temperature gradient, as well as the heat flux density. Each change is reflected by a separate lag value: a temperature time lag (τ_T) and a heat flux time lag (τ_q), respectively. The mathematical description of analyzed model can be presented in the form of the following system of equations:

$$\begin{cases} \nabla \circ \mathbf{q}(\mathbf{x},t) = -c_v \dfrac{\partial T(\mathbf{x},t)}{\partial t} + q_V(\mathbf{x},t) \\ k\tau_T \dfrac{\partial \nabla T(\mathbf{x},t)}{\partial t} + \tau_q \dfrac{\partial \mathbf{q}(\mathbf{x},t)}{\partial t} = -k \cdot \nabla T(\mathbf{x},t) - \mathbf{q}(\mathbf{x},t) \end{cases} \quad \mathbf{x} \in R^n,\ n \in N,\ t \in R_+ \cup \{0\} \tag{2}$$

The other general approach can be derived by using the General Equation for Non-Equilibrium Reversible-Irreversible Coupling (GENERIC) equation proposed in 1997 by M. Grmela and H. C. Öttinger [14–16]. This approach can be useful to obtain a model for the Monte Carlo simulation of modern nanostructures [17] and also for deterministic approaches [18]. Apart from that, an approach known as Guyer-Krumhansl-type heat conduction [19,20] can be also used. Moreover, the research presented by Pop et al. [21], related to heat generation and transportation problems in nanosized transistors, is also worth considering.

Let us consider, more precisely, one of the most common heat transfer models for electronic structures—the DPL model (Table 1a—left column), the ballistic-conductive heat transfer model (Table 1b—left column), and the model proposed in this paper, the DPL model with fractional order of the temperature function space derivative based on the Grünwald–Letnikov theory (Table 1c—left column). Suppose also an isotropic medium parameter properties and idempotent equations parameters $k > 0$, $c_v > 0$, k_{21}, k_{12}, τ_T, τ_q, τ_Q, etc. To simplify the analysis, the Taylor-series expansions mapping into a function with time delay is considered, as presented in Equation (3). The terms of higher orders are neglected.

$$f(\mathbf{x},t) + \tau \frac{\partial f(\mathbf{x},t+\tau)}{\partial t} \rightarrow f(\mathbf{x},t+\tau) \tag{3}$$

Then, time delayed PDEs are obtained and presented in the right column of Table 1. In the first case (Table 1a—right column), the state is appointed by using a gradient of a heat flux ($-k\cdot\Delta T$) delayed by τ_T-τ_q. It should be emphasized that results produced by the DPL model are consistent with many experiments and measurements at nanoscale (for more, see [9]).

In the case of the ballistic-conductive heat transfer model (Table 1b), the rate of value temperature $T(\mathbf{x}, t + \tau_q)$ change at $t+ \tau_q$ and $\mathbf{x}$ is dependent on a gradient of heat flux $(-k \cdot \Delta T)$ at the current time (t), increased by a difference of growth rate of the averaged value of temperature (T) over the infinitesimal neighborhood of point $\mathbf{x}$ and the value of temperature (T) in this point, both from the past time $(t - \tau_Q)$:

$$DT\left(\mathbf{x}, t + \tau_q\right) \propto k \cdot \Delta T(\mathbf{x}, t) \cdot Dt + |c_v k_{12} k_{21}| \cdot D\Delta T\left(\mathbf{x}, t - \tau_Q\right) \tag{4}$$

where $D(\cdot)$ is the difference operator ("change in") corresponding to changes for $Dt \to 0$ and also for $k_{12} \cdot k_{21} \leq 0$. Therefore, the temperature $T(\mathbf{x}, t + \tau_q)$ dynamic is additionally intensified by the dynamics of the rate of heat flux gradient $(\partial \Delta T(\mathbf{x}, t - \tau_Q)/\partial t)$ in the past time $(t - \tau_Q)$.

The approach proposed in this work is based on the temperature $T(\mathbf{x}, t + \tau_q)$ dynamic control, using the fractional order of Laplace operator $(_{GL}\Delta^{\alpha x})$ in the DPL model (Table 1c—right column). The integral or differential behavior of this operator and time–space discretization scheme is obtained by the value of parameter α_x. The application of the fractional derivative can be interpreted as an inhomogeneous space for the heat transfer in relation to the local energy distribution (interpreted as temperature at nanoscale) in infinitesimal neighborhood of considered point. Therefore, the infinitesimal distance in Laplacian is modulated in relation to the temperature distribution around the considered point at nanoscale. The physical interpretation of the fractional derivatives is also presented in [22–25].

Table 1. Selected Heat Transfer Model transformed into time delayed Partial Differential Equation.

Heat Transfer Model	Derived Time Delayed PDE
(a)	
DPL equation [12,13] for $\tau_T > \tau_q > 0$, Maxwell–Cattaneo–Vernotte equation [26–28] for $\tau_q > 0$, $\tau_T = 0$ (more details in Vermeersch and De Mey as well as Kovács and Ván papers [29,30]), and F–K Equations (1)–(2) for $\tau_q = \tau_T = 0$: $$\begin{cases} c_v \dfrac{\partial T(\mathbf{x},t)}{\partial t} = -\nabla \circ \mathbf{q}(\mathbf{x}, t) \\ \mathbf{q}(\mathbf{x}, t) + \tau_q \dfrac{\partial \mathbf{q}(\mathbf{x},t)}{\partial t} = -k \cdot \nabla T(\mathbf{x}, t) - k\tau_T \dfrac{\partial \nabla T(\mathbf{x},t)}{\partial t} \end{cases}$$	$$c_v \dfrac{\partial T\left(\mathbf{x},t+\tau_q\right)}{\partial t} = k \cdot \Delta T(\mathbf{x}, t + \tau_T)$$ $$\Downarrow$$ $$c_v \dfrac{\partial T\left(\mathbf{x},t+\tau_q-\tau_T\right)}{\partial t} = k \cdot \Delta T(\mathbf{x}, t)$$
(b)	
Ballistic-conductive heat transfer model for $\tau_q > 0$, $\tau_Q > 0$, $k_{12} \cdot k_{12} \leq 0$ (more details [30]) $$\begin{cases} c_v \dfrac{\partial T(\mathbf{x},t)}{\partial t} = -\nabla \circ \mathbf{q}(\mathbf{x}, t) \\ \tau_q \dfrac{\partial \mathbf{q}(\mathbf{x},t)}{\partial t} + \mathbf{q}(\mathbf{x}, t) = -k \cdot \nabla T(\mathbf{x}, t) - k_{21} \cdot \nabla Q(\mathbf{x}, t) \\ \tau_Q \dfrac{\partial Q(\mathbf{x},t)}{\partial t} + Q(\mathbf{x}, t) = k_{12} \cdot \nabla \circ \mathbf{q}(\mathbf{x}, t) \end{cases}$$	$$c_v \dfrac{\partial T\left(\mathbf{x},t+\tau_q\right)}{\partial t} =$$ $$= k \cdot \Delta T(\mathbf{x}, t) - c_v k_{12} k_{21} \dfrac{\partial \Delta T(\mathbf{x},t-\tau_Q)}{\partial t}$$
(c)	
Proposed DPL model [12,13], with fractional order of the temperature function space derivative based on Grünwald–Letnikov theory for $\tau_q > 0$, $\tau_T > 0$, $2 < \alpha_x < 2.5$ (more details in [31]): $$c_v \dfrac{\partial T(\mathbf{x},t)}{\partial t} + \tau_q c_v \dfrac{\partial^2 T(\mathbf{x},t)}{\partial t^2} =$$ $$= k \cdot {}_{GL}\Delta^{\alpha_x} T(\mathbf{x}, t) + k\tau_T \dfrac{\partial_{GL}\Delta^{\alpha_x} T(\mathbf{x},t)}{\partial t}$$	$$c_v \dfrac{\partial T\left(\mathbf{x},t+\tau_q\right)}{\partial t} = k \cdot {}_{GL}\Delta^{\alpha_x} T(\mathbf{x}, t + \tau_T)$$ $$\Downarrow$$ $$c_v \dfrac{\partial T\left(\mathbf{x},t+\tau_q-\tau_T\right)}{\partial t} = k \cdot {}_{GL}\Delta^{\alpha_x} T(\mathbf{x}, t)$$

Hence, the proposed model (and proposed time–space scheme) is a link between experimentally confirmed DPL mesoscopic model with the ballistic heat transport model with dynamic temperature changes' intensification useful for quasi 1-D nanostructures and for radiative heat transport without phonon collisions (e.g., metal nanowires and ultra-thin metal–oxide films).

One of the biggest advantages of the DPL approach is a universality of its application. It can be used in the case of parabolic partial differential equations, as well as for hyperbolic ones. It means that the Dual-Phase-Lag model can successfully replace the classical model based on parabolic

Fourier–Kirchhoff differential equation. Of course, the DPL approach also has some disadvantages. For example, it is characterized by a bigger complexity than F–K model, what results in longer time needed for computational simulation results, especially in the case of complex structures. However, this issue will not be investigated in this paper.

The crucial achievement described in this paper is the application of the Grünwald–Letnikov fractional space-derivative in Dual-Phase-Lag heat transfer model to more accurate modeling of delays and modulation of Laplacian in relation to the temperature distribution around the considered point. This approach seems to be beneficial to represent real heat transfer behavior at the nanoscale. The proposed solution is more accurate than the F–K heat transfer model. The computation time reduction is also obtained in comparison to the DPL model. Mentioned features will be demonstrated by using a special test nanostructure developed in microelectromechanical system (MEMS) technology.

The structure of the paper is as follows. Primarily, a description of investigated real test structure is prepared. Then, analyses related to mathematical modeling of the temperature distribution inside the test structure are included. The structure's discretization method is considered, as well. Key components of authors' approach, i.e., DPL model approximation scheme and heat transfer enhancement (including Grünwald–Letnikov fractional derivative and its application), are described in Sections 2.3 and 2.4. Next, simulation results are demonstrated and compared to real measurements described in [32,33]. Finally, results are discussed, and the research is summarized.

1.2. MEMS Test Structure Description

The MEMS test nanostructures were manufactured in the Polish Institute of Electron Technology. The structure consists of two parallel platinum resistors, which have lengths equal to 10 μm. The cross-sectional area of each resistor is a 100 nm square. Moreover, the distance between both resistors is also equal to 100 nm. They are placed on a thin layer of silicon dioxide, with a thickness of 100 nm. All layers are stacked on a 0.5 mm thick silicon wafer. A more detailed structure description can be found in Janicki et al. [33].

The test structure was bonded to a metal-core PCB characterized by high thermal conductivity. Moreover, the structure was connected to a biasing circuit, using high-frequency coaxial cables that were mounted to a tiny Hirose U.FL connector attached to the silicon die. Real photos of the analyzed structure, as well as the control circuit, are included in Janicki et al. [32,33].

During the measurement process, one of platinum resistors played a role of a heater, while the second one served as a temperature sensor. A detailed description of the measurement process of the test structure and obtained results is presented in [32,33].

2. Mathematical Description of Proposed Methodology

2.1. General Description

Thermal simulation was performed for the two-dimensional cross-sectional area of the investigated structure in the middle of the resistors' length. To obtain the temperature simulation results, the DPL model, Equation (2), was used. In order to make the analysis more effective, the system of Equation (2) was transformed to an equivalent form, presented below, for two-dimensional space [34]:

$$c_v \cdot \tau_q \cdot \frac{\partial^2 T(x,y,t)}{\partial t^2} + c_v \cdot \frac{\partial T(x,y,t)}{\partial t} - k \cdot \tau_T \cdot \frac{\partial \Delta T(x,y,t)}{\partial t} - k \cdot \Delta T(x,y,t) = q_V(x,y,t)$$
$$x,y \in R, \ t \in R_+ \cup \{0\} \tag{5}$$

The Laplacian of a temperature function ΔT was approximated by using the Finite Difference Method (FDM) according to the following formulas:

$$\Delta T(x,y,t) = \frac{\partial^2 T(x,y,t)}{\partial x^2} + \frac{\partial^2 T(x,y,t)}{\partial y^2}, \qquad x,y \in R, \ t \in R_+ \cup \{0\} \tag{6}$$

$$\frac{\partial^2 T(x,y,t)}{\partial x^2} \approx \frac{T(x+\Delta x,y,t) - 2\cdot T(x,y,t) + T(x-\Delta x,y,t)}{(\Delta x)^2}, x,y \in R,\ t \in R_+ \cup \{0\} \qquad (7)$$

$$\frac{\partial^2 T(x,y,t)}{\partial y^2} \approx \frac{T(x,y+\Delta y,t) - 2\cdot T(x,y,t) + T(x,y-\Delta y,t)}{(\Delta y)^2}, x,y \in R,\ t \in R_+ \cup \{0\} \qquad (8)$$

Thus, considering the same difference between nodes in both axes, i.e., $\Delta x = \Delta y$, Laplacian ΔT can be approximated in the following way:

$$\Delta T(x,y,t) \approx \frac{T(x+\Delta x,y,t)+T(x,y+\Delta x,t)-4\cdot T(x,y,t)+T(x-\Delta x,y,t)+T(x,y-\Delta x,t)}{(\Delta x)^2},$$
$$\text{dla } x,y \in R,\ t \in R_+ \cup \{0\} \qquad (9)$$

On the basis of this methodology, the authors' method for structure discretization and FDM matrices generation was proposed. Moreover, taking into consideration the proposed approximation, the DPL equation, in the form of Equation (5), has become an ordinary differential equation of a time variable. Finally, the prepared matrix system of equations was solved for different points of time, using a class of Backward Differentiation Formulas (BDF) [35–38].

In the following subsections, the structure discretization, as well as the proposed discretization scheme for DPL model describing the temperature distribution in the cross-sectional area of investigated test structure, is characterized in detail.

2.2. Structure Cross-Sectional Area Discretization

Primarily, the investigated cross-sectional area of the structure, as presented in Figure 1, was discretized by using two-dimensional discretization mesh characterized by the following formulas [34]:

$$\mathbf{q}_k(t) = \mathbf{q}(x,y,t), \qquad x = i\cdot\Delta x,\ y = j\cdot\Delta y \qquad (10)$$

$$T_k(t) = T(x,y,t), \qquad x = i\cdot\Delta x,\ y = j\cdot\Delta y \qquad (11)$$

$$i \in \{1,2,\dots,n_x\}, j \in \{1,2,\dots,n_y\}, k \in \{1,2,\dots,n_x \cdot n_y\}$$

where n_x and n_y describe a number of discretization nodes in the x-axis and y-axis, respectively. On the other hand, the product $n_x \cdot n_y$ reflects the entire number of nodes used to discretize the structure's cross-section.

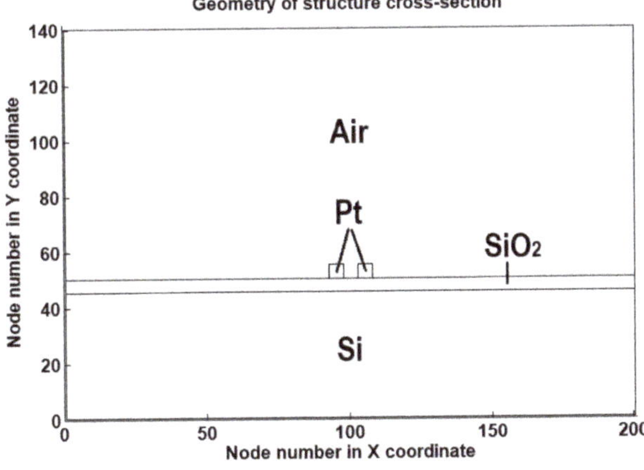

Figure 1. Geometry of cross-sectional area of the investigated test structure.

Nodes are numbered from the left to the right side, along the *x*-axis. After reaching the last point in a single row, the next part of the structure, being the nearest row from the top of the current row, was taken into consideration and numbered in the same way. Thus, node no. 1 was placed in the left bottom corner of the structure, while the node with the highest possible number, equal to $n_x \cdot n_y$, was located in the top right corner. The graphical representation of used discretization mesh for analyzed cross-section of the test structure, as well as the way mesh nodes were numbered, is demonstrated in Figure 2. It is also worth highlighted that the distance between nodes in both dimensions was set to 10 nm.

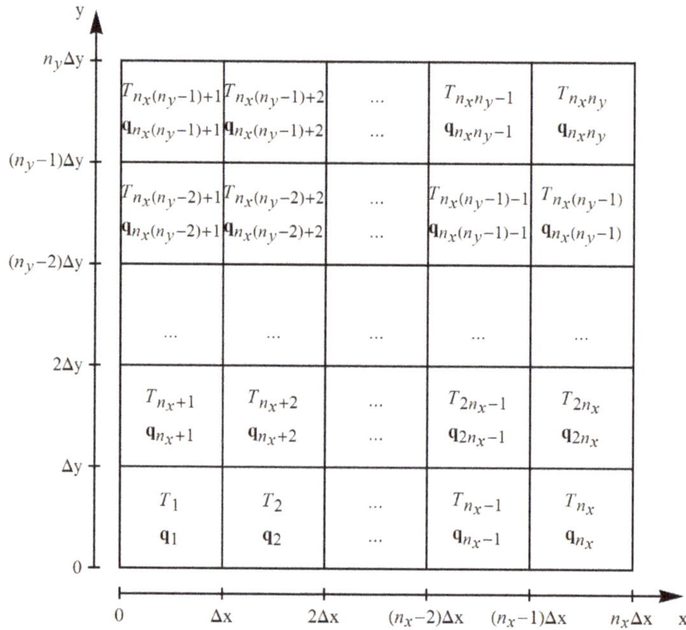

Figure 2. The graphical representation of discretization mesh and nodes' numbering. Reprint with permission [34,39]; Copyright 2018, ŁTN.

2.3. Dual-Phase-Lag Approximation Scheme for Test Structure

In order to obtain the solution of temperature distribution problem inside the investigated cross-sectional area of the structure, the Dirichlet boundary conditions were considered at the bottom, while left, right and the top parts have been modeled using Neumann boundary conditions. Described situation can be characterized by the following equations:

$$T_k(t) = 0, \quad t \in R_+ \cup \{0\}, \\ k \in \{1, 2, 3, \ldots, n_x\} \tag{12}$$

$$q_k(t) = 0, \quad t \in R_+ \cup \{0\}, \\ k \in \left\{n_x + 1, 2 \cdot n_x + 1, \ldots, (n_y - 1) \cdot n_x + 1\right\} \cup \left\{n_x, 2 \cdot n_x, \ldots, (n_y - 1) \cdot n_x\right\} \cup \\ \cup \left\{(n_y - 1) \cdot n_x + 1, (n_y - 1) \cdot n_x + 2, \ldots, n_y \cdot n_x\right\} \tag{13}$$

Considering imposed boundary conditions and the FDM assumptions, the DPL model can be explained by the following system:

$$\begin{cases} \mathbf{M} \cdot \ddot{\mathbf{T}}(t) + \mathbf{D} \cdot \dot{\mathbf{T}}(t) + \mathbf{K} \cdot \mathbf{T}(t) = \mathbf{b} \cdot u(t) \\ \qquad \mathbf{y}(t) = \mathbf{c}^T \cdot \mathbf{T}(t) \end{cases} \qquad t \in R_+ \cup \{0\} \qquad (14)$$

where index T means a transposition, while $\mathbf{T}$, $\dot{\mathbf{T}}$, and $\ddot{\mathbf{T}}$ are $n_x{\cdot}n_y$-element vectors reflecting the temperature function and its first and the second time derivatives, respectively. Moreover, taking into account the way of nodes numbering, matrices $\mathbf{I_{diag}}$, $\mathbf{M_{FDM}}$, $\mathbf{M}$, $\mathbf{D}$, $\mathbf{K}$, and $\mathbf{c}^T$; vectors $\mathbf{c_v}$, $\mathbf{k}$, $\boldsymbol{\tau_q}$, $\boldsymbol{\tau_T}$, $\mathbf{b}$, and $\mathbf{y}$; and the function $u(t)$ may be reflected as indicated below:

$$\mathbf{I_{diag}}, \mathbf{M_{FDM}}, \mathbf{M}, \mathbf{D}, \mathbf{K}, \mathbf{c}^T \in R^{n_x{\cdot}n_y \times n_x{\cdot}n_y},$$
$$\mathbf{c_v}, \mathbf{k}, \boldsymbol{\tau_q}, \boldsymbol{\tau_T}\, \mathbf{b}, \mathbf{y} \in R^{n_x{\cdot}n_y \times 1},$$
$$u \in R$$

$$\mathbf{M_{FDM}} = \begin{bmatrix} -3 & 1 & 0 & \ddots & 1 & 0 & 0 & \ddots & 0 & 0 & 0 \\ 1 & -4 & 1 & \ddots & 0 & 1 & 0 & \ddots & 0 & 0 & 0 \\ 0 & 1 & -4 & \ddots & 0 & 0 & 1 & \ddots & 0 & 0 & 0 \\ \ddots & \ddots & \ddots & \ddots & \ddots & \ddots & \ddots & \ddots & \ddots & \ddots & \ddots \\ 1 & 0 & 0 & \ddots & -4 & 1 & 0 & \ddots & 1 & 0 & 0 \\ 0 & 1 & 0 & \ddots & 1 & -4 & 1 & \ddots & 0 & 1 & 0 \\ 0 & 0 & 1 & \ddots & 0 & 1 & -4 & \ddots & 0 & 0 & 1 \\ \ddots & \ddots & \ddots & \ddots & \ddots & \ddots & \ddots & \ddots & \ddots & \ddots & \ddots \\ 0 & 0 & 0 & \ddots & 1 & 0 & 0 & \ddots & -3 & 1 & 0 \\ 0 & 0 & 0 & \ddots & 0 & 1 & 0 & \ddots & 1 & -3 & 1 \\ 0 & 0 & 0 & \ddots & 0 & 0 & 1 & \ddots & 0 & 1 & -2 \end{bmatrix} \qquad (15)$$

$$\mathbf{M} = \mathrm{diag}(\boldsymbol{\tau_q}) \circ \mathrm{diag}(\mathbf{c_v}) \circ \mathbf{I_{diag}} \qquad (16)$$

$$\mathbf{D} = \mathrm{diag}(\mathbf{c_v}) \circ \mathbf{I_{diag}} - \frac{1}{(\Delta x)^2} \cdot \mathrm{repmat}(\boldsymbol{\tau_T}) \circ \mathrm{repmat}(\mathbf{k}) \circ \mathbf{M_{FDM}} \qquad (17)$$

$$\mathbf{K} = -\frac{1}{(\Delta x)^2} \mathrm{repmat}(\mathbf{k}) \circ \mathbf{M_{FDM}} \qquad (18)$$

$$\mathbf{b} = a \cdot \begin{bmatrix} 0 & \cdots & 0 & 1 & \cdots & 1 & 0 & \cdots & 0 \end{bmatrix}^T, \qquad a \in R_+ \qquad (19)$$

$$\mathbf{c} = \mathbf{I_{diag}} \qquad (20)$$

where $\mathbf{I_{diag}}$ is an identity matrix, operator $\circ$ reflects matrices multiplication resulting in a Hadamard product, and matrix function diag $(\cdot)$ creates a diagonal matrix from a vector, while Matlab repmat function replicates a given vector and composes a matrix of required dimensions. It is also worth highlighting that non-zero values in vector $\mathbf{b}$ are observed only in the case of mesh nodes characterizing a heating area.

The system in Equation (14) of differential equations of the second order was equivalently transformed into the following system of the linear first-order equations [39]:

$$\begin{cases} \mathbf{E} \cdot \dot{\overline{\mathbf{T}}}(t) = \mathbf{A} \cdot \overline{\mathbf{T}}(t) + \mathbf{B} \cdot u(t) \\ \mathbf{y}(t) = \mathbf{c}^T \cdot \overline{\mathbf{T}}(t) \end{cases} \qquad t \in R_+ \cup \{0\} \tag{21}$$

where $\overline{\mathbf{T}}$ and $\dot{\overline{\mathbf{T}}}$ are $2 \cdot n_x \cdot n_y$-element vectors. The first part of vector $\overline{\mathbf{T}}$ consists of elements of vector $\mathbf{T}$, while its second part includes elements of $\dot{\mathbf{T}}$. On the other hand, the first and the second parts of the vector $\dot{\overline{\mathbf{T}}}$ consists of elements of vectors $\dot{\mathbf{T}}$ and $\ddot{\mathbf{T}}$, respectively. Moreover, matrices $\mathbf{E}$, $\mathbf{A}$, $\mathbf{B}$, and $\mathbf{C}^T$ can be represented in the following way [39,40]:

$$\mathbf{E} = \begin{bmatrix} \mathbf{I}_{diag} & \Theta \\ \Theta & \mathbf{M} \end{bmatrix}, \quad \mathbf{A} = \begin{bmatrix} \Theta & \mathbf{I}_{diag} \\ -\mathbf{K} & -\mathbf{D} \end{bmatrix}, \quad \mathbf{B} = \begin{bmatrix} \Theta_1 \\ \mathbf{b} \end{bmatrix}, \quad \mathbf{C} = \begin{bmatrix} \mathbf{c} & \Theta \\ \Theta & \mathbf{c} \end{bmatrix} \tag{22}$$

where Θ and Θ_1 are null matrices. Moreover, we get the following:

$$\mathbf{E}, \mathbf{A}, \mathbf{C}^T \in R^{2 \cdot n_x \cdot n_y \times 2 \cdot n_x \cdot n_y}, \ \mathbf{B} \in R^{2 \cdot n_x \cdot n_y \times 1}, \ \mathbf{I}, \Theta \in R^{n_x \cdot n_y \times n_x \cdot n_y}, \ \Theta_1 \in R^{n_x \cdot n_y \times 1}$$

Thus, the full authors' approximation scheme for the DPL model in two-dimensional Euclidean space is described by the system in Equation (21), including explanations formulated in Equations (15)–(22) and boundary conditions in Equations (12) and (13). The final solution of temperature-distribution changes over time in the analyzed cross-section of the test structure, based on the prepared approximation scheme, is determined by using the BDF method of a variable order between 1 and 5.

2.4. Heat Transfer Enhancement

The considered test structure was analyzed, including the surrounding air environment. Furthermore, platinum resistors' separation distance $d = 100$ nm in the benchmark structure is comparable to the surrounded air mean free path length $\Lambda \approx 65$ nm. Therefore, the heat flow can be approximated by the following equation [41]:

$$q_{air} = \frac{k_{air}}{\langle d + a\Lambda \rangle}(T_{Pt,A} - T_{Pt,B}) \tag{23}$$

where a is used to describe the interaction between the gas molecules and the solid walls [42], typically $a = 1$, and $<\cdot>$ stands for ensemble average. The final equivalent air conductance between the mentioned resistors was estimated by using the following simplified formula (compare with value in Table 2):

$$k_{air,eqv,PtA-PtB} = k_{air}\frac{d}{\langle d + a\Lambda \rangle} \approx 0.961 \, k_{air} \approx 0.961 \cdot 0.0263 \, \text{W}/(\text{mK}) \tag{24}$$

The photon tunneling and the radiative heat transfer between platinum resistor parallel surfaces were also analyzed. The photon tunneling was neglected due to the small amount of heat flux transport $(S_z)_{max}^{evamescant} = 2.2 \cdot 10^{-19} \text{W}/\text{m}^2$ (calculated for vacuum environment) in comparison to the main heat flux Pt-SiO$_2$ $q_{Pt\_SiO2,cond} \approx 19.3$ MW/m^2 and the heat conduction through $q_{Pt\_Pt,cond} \approx 0.917 \, q_{Pt\text{-}SiO2,cond} >> (S_z)_{max}^{evamescant}$ in the air for the steady state [43,44]:

$$(S_z)_{max}^{evamescant} \approx \frac{k_B^2 \cdot (T_{PtA} + 273.15\text{K})}{24\hbar b^2} \left[\text{W}/\text{m}^2\right] \tag{25}$$

where b is an inter-atomic distance $b \approx 39.12$ nm for Pt, k_B means the Boltzmann constant, and $\hbar$ is the reduced Planck constant. Moreover, the radiative heat transfer ($q_{SB\ rad}$) was also neglected, due to the insignificant emitted energy from warmer (T_{PtA}) to colder platinum resistor (T_{PtB}) parallel surfaces $q_{Pt\text{-}SiO2,\ cond}\ /\ q_{SB\ rad} \approx 8.3 \cdot 10^{25}$:

$$q_{SB\ rad} \approx 5.6693\ \frac{W}{m^2\left(\frac{K}{100}\right)} \cdot \left(\sqrt{2}-1\right) \cdot \left\{\left(\frac{T_{PtA}+273.15K}{100}\right)^4 - \left(\frac{T_{PtB}+273.15K}{100}\right)^4\right\} \tag{26}$$

$$\text{and } 72 \le q_{SB\ rad} \le 146W/m^2 \text{ for } T_{PtA} \approx 59K \text{ and } 4.7K \le T_{PtB} \le 31K$$

Moreover, considering investigation presented above, some additional changes in the proposed model are needed. Thus, for the air area between resistors' surfaces and for contact areas between platinum and silicon dioxide, a fractional order of the temperature-rise function space derivative has been employed, according to the theory described in [31,45]. This theory, based on the Grünwald–Letnikov definition of the fractional derivative, allows us to establish the following formula reflecting the temperature-rise function's improvement for the central difference of the FDM scheme [31,45]:

$$_{GL}D_{0,s}^{\alpha_x}T(s) = \frac{1}{(\Delta s)^\alpha} \cdot \sum_{k=0}^{\text{round}(\alpha_x,0)} (-1)^k \frac{\Gamma(\alpha_x+1)}{\Gamma(k+1)\cdot\Gamma(\alpha_x-k+1)} T\left(s - k\cdot\Delta s + \frac{\alpha_x\cdot\Delta s}{2}\right), \tag{27}$$

$$\text{for } \alpha_x \in R_+,\ \Delta s \to 0$$

where Δs is the mesh nodes distance, α is investigated fractional order, Γ is the special gamma function, and $round(m,n)$ is the function rounding the value m to n digits. Considering fractional order of derivative, we needed approximating investigated function values in points between nodes of a discretization mesh. The mentioned approximation depends on function values in neighboring mesh points. Thus, the right-hand-side part of Equation (27), being an investigated approximation, can be reflected by using the following equations [45]:

$$\frac{1}{(\Delta s)^{\alpha_x}} \cdot \sum_{k=0}^{2} (-1)^k \frac{\Gamma(\alpha_x+1)}{\Gamma(k+1)\cdot\Gamma(\alpha_x-k+1)} T\left(\mathbf{x} - k\cdot\Delta\mathbf{x} + \frac{\alpha_x\cdot\Delta\mathbf{x}}{2}, t\right) \cdot 1 =$$

$$= \frac{\left(\frac{\alpha_x}{2}-1\right)\cdot T(x+2\cdot\Delta x,y,t) + \left(2-\frac{\alpha_x}{2}-\alpha_x\cdot\left(\frac{\alpha_x}{2}-1\right)\right)\cdot T(x+\Delta x,y,t)}{(\Delta x)^{\alpha_x}} +$$

$$+ \frac{\left(\frac{\alpha_x\cdot(\alpha_x-1)}{2}\cdot\left(\frac{\alpha_x}{2}-1\right)-\alpha_x\cdot\left(2-\frac{\alpha_x}{2}\right)\right)\cdot T(x,y,t) + \left(\left(2-\frac{\alpha_x}{2}\right)\cdot\frac{\alpha_x\cdot(\alpha_x-1)}{2}\right)\cdot T(x-\Delta x,y,t)}{(\Delta x)^{\alpha_x}}$$

$$+ \frac{\left(\frac{\alpha_x}{2}-1\right)\cdot T(x,y+2\cdot\Delta x,t) + \left(2-\frac{\alpha_x}{2}-\alpha_x\cdot\left(\frac{\alpha_x}{2}-1\right)\right)\cdot T(x,y+\Delta x,t)}{(\Delta x)^{\alpha_x}} +$$

$$+ \frac{\left(\frac{\alpha_x\cdot(\alpha_x-1)}{2}\cdot\left(\frac{\alpha_x}{2}-1\right)-\alpha_x\cdot\left(2-\frac{\alpha_x}{2}\right)\right)\cdot T(x,y,t) + \left(\left(2-\frac{\alpha_x}{2}\right)\cdot\frac{\alpha_x\cdot(\alpha_x-1)}{2}\right)\cdot T(x,y-\Delta x,t)}{(\Delta x)^{\alpha_x}} \tag{28}$$

$$\text{for } \alpha_x \in (2,\ 2.5),\ \Delta x \to 0$$

Table 2. Considered material parameters' values [46].

Layer	Material	$k\left[\frac{W}{m\cdot K}\right]$	$\rho\left[\frac{kg}{m^3}\right]$	$c_p\left[\frac{J}{kg\cdot K}\right]$
1 (wafer)	Silicon (Si)	148	2330	712
2 (oxide)	Silicon dioxide (SiO$_2$)	1.38	2220	745
3 (heater)	Platinum (Pt)	71.6	21,450	133
4 (thermometer)	Platinum (Pt)	71.6	21,450	133
5 (ambient)	Air	0.0263 *	1.1614	1.007

* See Section 2.4.

The formula above replaces the classic approximation of the Laplacian described by Equation (9).

3. Thermal Simulation and Results Analysis

3.1. Material Characterization and Initial Simulation Results

A thermal simulation of the test structure was prepared by using MathWorks®Matlab environment and proposed author's approximation scheme for the DPL model. The used computational node includes 4-core, 8-threads Intel® Core™ i7 2.6 GHz (3.6 GHz in Turbo mode) CPU, 32 GB DDR4 memory supported by 265 GB of swap file. In order to obtain simulation results, parameters' values presented in Table 2 were taken into consideration.

The most problematic issue is related to establishing parameters of the air layer, being an ambient of investigated test structure. In particular, the air between two platinum resistors is crucial in presented investigation. Moreover, the contact layer between platinum resistors and the oxide layer also needs special consideration. In order to emphasize the problem, a sample analysis of temperature rises over the time were prepared for different values of the DPL model parameters, as well as for different values of a thermal conductivity for the mentioned part of the air layer.

The first part of the simulation process does not include the investigation demonstrated in Section 2.4. Two sets of reference DPL model parameters were considered during the simulation process:

- $\tau_T = 60$ ps, $\tau_q = 3$ ps (all layers)
- $\tau_T = 2.6$ ps, $\tau_q = 0.0916$ ps (platinum resistors) and $\tau_T = 60$ ps, $\tau_q = 3$ ps (all remaining layers).

The results, as demonstrated in Figures 3 and 4, were additionally normalized in order to make their analysis easier. The normalization was prepared in the following way:

$$T_k^{norm}(t) = \frac{T_k(t)}{\max\limits_{t,k}\{T_k(t)\}} \qquad k \in \left\{1, 2, \ldots, n_x \cdot n_y\right\}, t \in R_+ \cup \{0\} \tag{29}$$

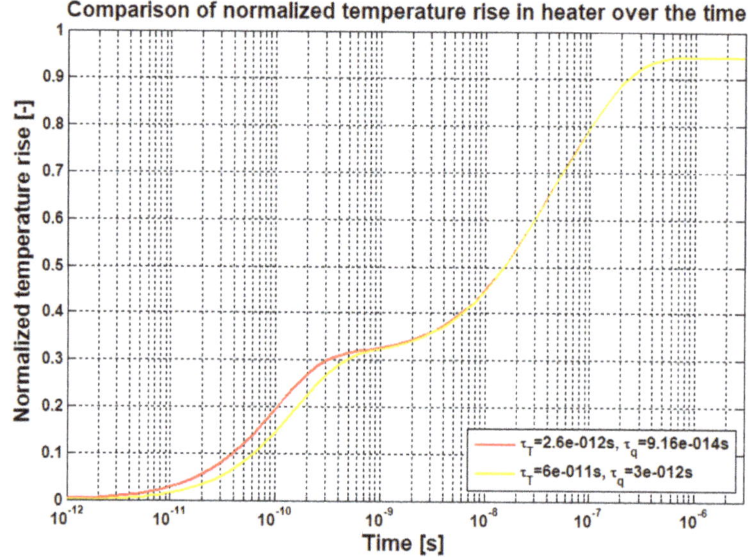

Figure 3. Average temperature rises over the time inside heat source (excluding investigation in Section 2.4).

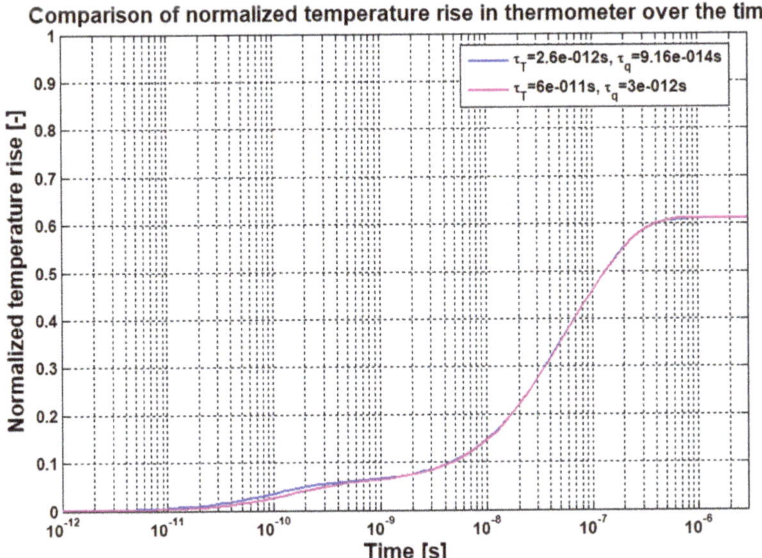

Figure 4. Average temperature rises over the time inside temperature sensor (excluding investigation in Section 2.4).

As it can be seen, the average temperature rise inside the heater is less than 95% of the maximal recorded temperature rise, while the temperature rise in the platinum sensor is characterized by a slightly more than 61% of the highest observed temperature rise. Differences in temperature rise values yielded for analyzed sets of DPL model parameters are observed between 1 ps and 1 ns.

Time shifts between observed lines in Figures 3 and 4 were plotted in Figure 5. In order to show differences between results, excluding and including air conductivity investigation presented in Section 2.4, a time shift analysis over the time, demonstrated in Figure 6, was carried out. In this figure, "new k_{air}" means the thermal conductivity of the air layer between platinum resistors calculated based on the analysis shown in Section 2.4. As a comparison, the simulation results obtained for both DPL time lags equal to zero for the air layer are also included.

Taking into consideration the sensitivity of yielded results to even small changes of the air layer material parameters, in the second part of the simulation, values, calculated considering the investigation described in Section 2.4, were employed only.

3.2. Final Simulation Results and Comparison to Real Measurements of the Test Structure

The analysis in the previous subsections shows that there is a need for calculation of the proper material parameters for the air layer, especially between platinum resistors of the test structure. Moreover, results of further research suggest using the fractional order of the space derivative of the temperature rise function for the investigated region, as well as this one between platinum resistors and the silicon dioxide, according to the theory described in Section 2.4. Taking into consideration these facts, the simulation of the temperature distribution in the cross-sectional area of the real test structure was carried out.

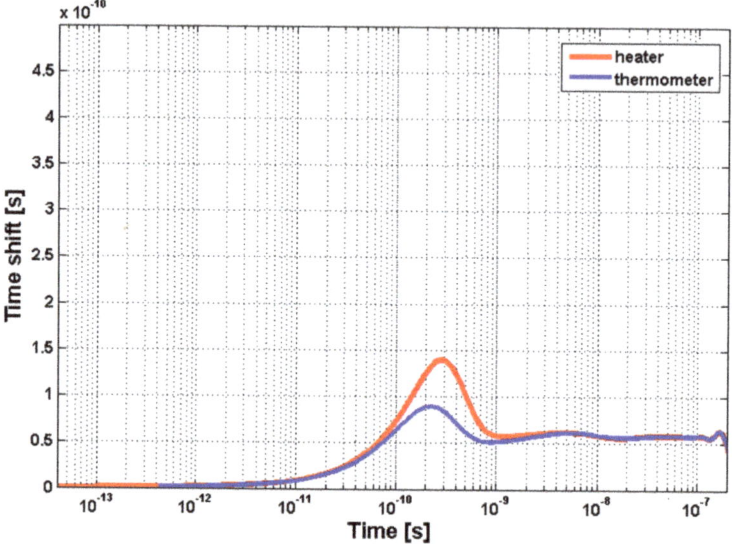

Figure 5. Time shifts over the time for temperature rises in heat source and temperature sensor (excluding investigation in Section 2.4).

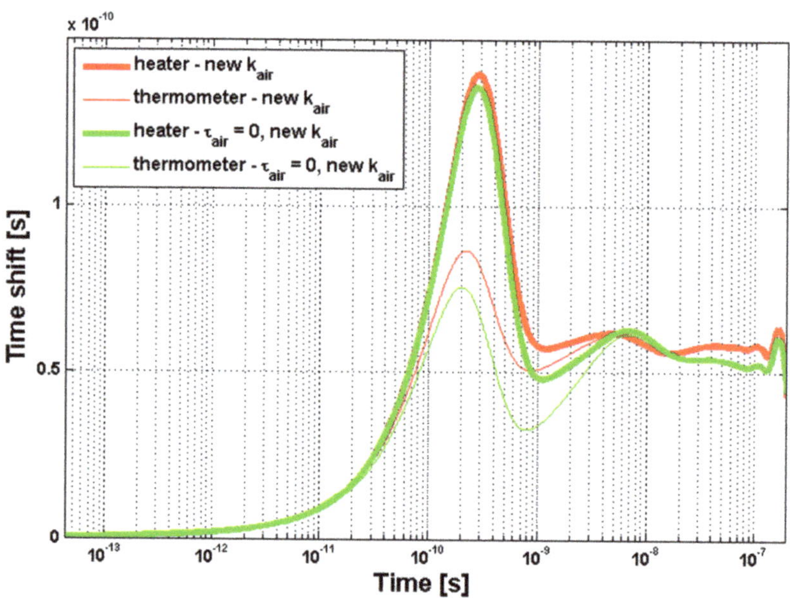

Figure 6. Time shifts over the time for temperature rises observed in the heat source and temperature sensor (including analysis in Section 2.4).

It was assumed that differences between mesh nodes are equal to 10 nm in both axes. Furthermore, values of DPL model time-lag parameters were calculated according to the theory presented in

Zubert et al. [8]. Thus, for Platinum resistors, heat-flux time lag was set at approximately 550 ps, while the considered value of the temperature time lag was equal to 15.6 ns. In the case of other materials, investigated parameters were equal to 18 and 480 ns, respectively. Simulation results and their comparison to the real measured data (collected and described in [32,33]) are presented in Figures 7 and 8, for the transient and steady state analyses, respectively. Moreover, in Figure 7, simulation results received by using the classical F–K model are also included, for comparison purposes.

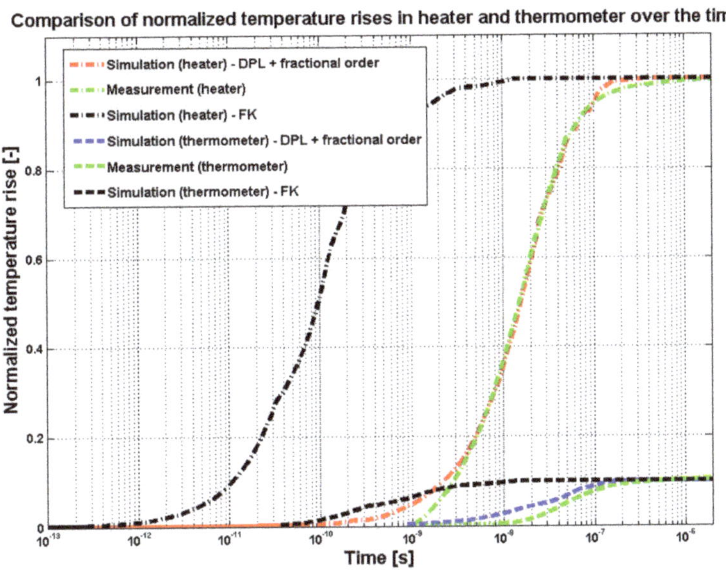

Figure 7. Comparison of normalized temperature rises in heater and thermometer.

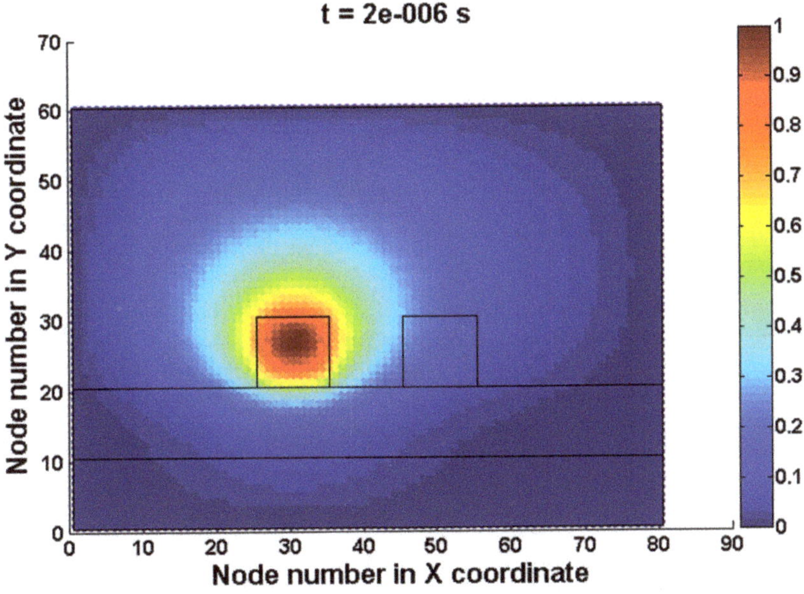

Figure 8. Steady-state temperature distribution in cross-sectional area of investigated structure.

In Figure 7, the results for the heated platinum resistor were marked by dashed and dotted lines, while those ones obtained for the temperature sensor were plotted with dashed lines. Moreover, outputs for the heater and thermometer were marked by the red and blue curves, respectively. On the other hand, measurement data (collected and described in Janicki et al. [32,33]) were plotted, using green lines, while results obtained by using the F–K model were marked by the black color.

The maximal temperature rise above the ambient temperature was observed at the top surface of the heat source, i.e., in the first platinum resistor (layer 3). Its value is nearly 60 K. On the other hand, the temperature rise recorded at the surface of the temperature sensor, i.e., the second platinum resistor (layer 4), is about 6 K, which states approximately 10% of the temperature rise observed in the heat source.

Both of the investigated temperature rises coincide almost exactly with measurements of the real structure (green curves in Figure 7). Parameter α_x allows for changing the result curves' slopes, while DPL model parameters τ_q and τ_T cause the curves to shift over the time. The mentioned change is proportional to the parameters' values.

In order to check the quality of the obtained results for used the discretization mesh and mesh nodes distances (10 nm), which were described in Section 2.2, the simulation curves' fitting to the measured one was considered. Each curve was plotted based on 703 points logarithmically distributed over the time. Then, such metrics as coefficient of determination (R2), root mean squared error (RMSE), sum of squared estimate error (SSE), and mean squared error (MSE) were calculated. Moreover, to assess the goodness of recognition as a volatility over the time, a Pearson correlation coefficient (corr) was also considered. Determined metrics values are presented in Table 3.

Table 3. Evaluation of goodness of fitting of simulation results to real data.

Temperature Distribution Simulation	MSE	RMSE	SSE	R^2	Corr
Heater	$8.5837 \cdot 10^{-4}$	0.0293	0.6034	0.9572	0.9973
Thermometer	$9.5540 \cdot 10^{-5}$	0.0098	0.0472	0.9554	0.9748

As it can be seen, both the heater and thermometer curves are characterized by relatively small values of MSE, RMSE, and SSE values. Moreover, the coefficient of determination and correlation coefficient suggest a proper recognition of a shape of the measured curve by a simulated one. Generally, it can be stated that calculated metrics (MSE, RMSE, SSE, coefficient of determination, and correlation coefficient) for the simulation fitting to the real data confirm highly accurate simulation results. This situation clearly shows that the proposed approach based on the theory described in Section 2 allows for the production of outputs reflecting the real thermal phenomena observed at the nanoscale. Moreover, as it was shown in Figure 7, the classical F–K model should not be used in the case of electronic nanosized structures.

4. Conclusions

This paper includes the investigation of the heat transfer problems at the nanoscale. A new approach to the heat transfer modeling in modern nanosized structures was considered. It combines the DPL model and the Grünwald–Letnikov fractional derivative. A combination of these mathematical tools allows for the preparation of a complex approach using the Finite Difference Method to temperature distribution determination at the nanoscale, with a high level of accuracy, as is confirmed by the measurement of a real test structure.

An important novelty described in this paper is the use of a DPL model with a fractional order space derivative of a temperature function based on the Grünwald–Letnikov derivative operator. This operator, as well as the proposed time–space discretization schema, is a bridge between experimentally confirmed DPL mesoscopic model with the ballistic heat transport model with dynamic temperature changes' intensification useful for quasi 1-D nanostructures and for radiative heat transport without phonon collisions. This solution allows for the consideration of such physical behaviors like time needed for a heat flux or temperature gradient changes. Thus, a modeling of a heat diffusion can be investigated by making realistic assumptions, which was not possible in the case of the F–K model use and real relaxation thermal properties of material at mesoscopic scale. The research has shown that the proposed GL DPL model is more realistic than the commonly used Fourier–Kirchhoff model.

The manuscript describes also proposed an approximation scheme of a modern DPL heat transfer model based on the Finite Difference Method approach prepared for the two-dimensional cross-section of the real test nanometric electronic structure manufactured at the Institute of Electron Technology in Warsaw. The investigation has shown that there is a possibility to effectively implement a prepared algorithm that allows for the determination of a temperature distribution inside real nanoscale electronic structures, based on proposed an approximation scheme.

Thermal simulation has provided results which coincide almost exactly with the real measurements. It means that prepared methodology is highly accurate and allows modeling of the heat transfer problems by using a modern approach based on the use of the Dual-Phase-Lag model. The considered thermal model is an appropriate methodology for heat-diffusion modeling, especially at the nanoscale.

In the future, the reduction of the DPL model order reduction methodology will be considered in order to save simulation time, decrease a computational power requirements [47], and make the simulation process more efficient.

Author Contributions: The algorithm for the FDM approximation scheme of DPL model for two-dimensional cross-section of investigated test structure, analysis of the Grünwald–Letnikov temperature derivative of a fractional order and the algorithm calculated its values, numerical simulations and evaluation of their results, and preparation of this manuscript were carried out by T.R. Preparation of the algorithm convergence analysis of numerical simulations, equivalent air conductance investigation, and photon tunneling, as well as the radiative heat transfer calculation, thermodynamic models aspects and time delayed PDEs analysis, were performed by M.Z.; M.Z. also supervised the research and made corrections to the manuscript. All authors have read and agreed to the published version of the manuscript.

Funding: The research presented in this paper was carried out and funded within the Polish National Science Centre project OPUS No. 2016/21/B/ST7/02247.

Acknowledgments: Authors would like to express their special thanks to M. Janicki and J. Topilko sharing papers [32,33].

Conflicts of Interest: The authors declare no conflicts of interest.

Nomenclature

Symbol	Description	SI Unit
	Latin symbols	
T	Temperature rise distribution in analyzed area in relations to ambient temperature	K
a	Constant describing an interaction between gas molecules and solid walls	-
b	Inter-atomic distance	m
c_p	Specific heat of a material for a constant pressure (c_p)	$\frac{J}{kg \cdot K}$
c_v	Volumetric heat capacity being a product of a specific heat of a material for a constant pressure (c_p) and its density (ρ)	$\frac{J}{m^3 \cdot K}$
$\hbar$	Quotient of Planck constant and the value of $2 \cdot \pi$	$J \cdot s$
k	Material thermal conductivity	$\frac{W}{m \cdot K}$
k_B	Boltzmann constant	$\frac{J}{K}$
q	Heat flux density	$\frac{W}{m^2}$
q_V	Volume density of internally generated heat	$\frac{W}{m^3}$
Δs	Mesh nodes distance	m
t	Time variable	s
x	Space variable $$\mathbf{x} = \begin{cases} [x_1, x_2, \ldots, x_n]^T, \mathbf{x} \in R^n, n \in N \setminus \{1\} \\ \quad x, \qquad \mathbf{x} \in R \end{cases}$$	m^n

Symbol	Description	SI Unit
	Greek symbols	
α_x	Order of a fractional Grünwald–Letnikov space derivative	-
Λ	Molecule's mean free path length	m
ρ	Material density	$\frac{kg}{m^3}$
τ_T	Temperature time lag	s
τ_q	Heat flux time lag	s
	Matrix and vectors	
1	Vector including only 1 values	-
$a \times b$	Matrix dimensions; a reflect number of rows, b is the number of columns	-
	Derivatives	
∂	Derivative symbol	-
f'	First derivative of function f	-
f''	Second derivative of function f	-
${}_{GL}D_{0,s}^{\alpha}$	Fractional Grünwald–Letnikov derivative of order α around point 0 for s variable	-
	Mathematical operators	
$D(\cdot)$	Difference operator corresponding to changes for $Dt \to 0$	-
	Multiplication operator	-
∇	Nabla operator	-
Δ	Laplace operator	-
${}_{GL}\Delta^{\alpha x}$	Fractional order of Laplace operator	-
$\nabla \circ$	Divergence operator in orthogonal Euclidean space	-
$\cup$	Sets' union operator	-
T	Transposition operator	-
	Sets and spaces	
$\{a_1, a_2, a_3, \ldots, a_n\}$	Finite set of elements	-
(a, b)	Open interval between a and b	-
span$\{$**a**,**b**$\}$	Linear subspace generated by vectors **a** and **b**	-
	Functions	
$\langle \cdot \rangle$	Ensemble average	-
$\lceil \alpha \rceil$	Rounding of α value to the smallest integer number higher or equal to α	-
diag$(\cdot)$	Matrix function creating a diagonal matrix from a vector	-
repmat$(\cdot)$	Matrix function replicating a given vector and composing a matrix of required dimensions	-
round(α,k)	Rounding of α value to k^{th} digit after decimal point	-
$\max_{s}\{f(s)\}$	Maximum of the set including f function values	-
$\Gamma(\cdot)$	Special Gamma function	-

References

1. Fourier, J.-B.J. *Théorie Analytique de la Chaleur*; Firmin Didot: Paris, France, 1822.
2. Fourier, J.-B.J. *The Analytical Theory of Heat*; Cambridge University Press: London, UK, 1878.
3. Raszkowski, T.; Samson, A. The Numerical Approaches to Heat Transfer Problem in Modern Electronic Structures. *Comput. Sci.* **2017**, *18*, 71–93. [CrossRef]
4. Raszkowski, T.; Zubert, M.; Janicki, M.; Napieralski, A. Numerical solution of 1-D DPL heat transfer equation. In Proceedings of the 22nd International Conference Mixed Design of Integrated Circuits and Systems (MIXDES), Torun, Poland, 25–27 June 2015; pp. 436–439.
5. Zubert, M.; Janicki, M.; Raszkowski, T.; Samson, A.; Nowak, P.S.; Pomorski, K. The Heat Transport in Nanoelectronic Devices and PDEs Translation into Hardware Description Languages. *Bulletin de la Société des Sciences et des Lettres de Łódź Série Recherches sur les Déformations* **2014**, *LXIV*, 69–80.
6. Nabovati, A.; Sellan, D.P.; Amon, C.H. On the lattice Boltzmann method for phonon transport. *J. Comput. Phys.* **2011**, *230*, 5864–5876. [CrossRef]
7. Zubert, M.; Raszkowski, T.; Samson, A.; Janicki, M.; Napieralski, A. The distributed thermal model of fin field effect transistor. *Microelectron. Reliab.* **2016**, *67*, 9–14. [CrossRef]
8. Zubert, M.; Raszkowski, T.; Samson, A.; Zając, P. Methodology of determining the applicability range of the DPL model to heat transfer in modern integrated circuits comprised of FinFETs. *Microelectron. Reliab.* **2018**, *91*, 139–153. [CrossRef]
9. Tzou, D.Y. *Macro-to Microscale Heat Transfer: The Lagging Behavior*, 2nd ed.; Willey: Hoboken, NJ, USA, 2015.

10. Jou, D.; Casas-Vázquez, J.; Lebon, G. Extended Irreversible Thermodynamics. *Rep. Prog. Phys.* **1988**, *51*, 1105–1179. [CrossRef]

11. Jou, D.; Casas-Vázquez, J.; Lebon, G. *Extended Irreversible Thermodynamics*, 4th ed.; Springer: Berlin/Heidelberg, Germany, 2009.

12. Tzou, D.Y. An Engineering Assessment to the Relaxation Time in Thermal Waves. *Int. J. Heat Mass Transf.* **1993**, *36*, 1845–1851. [CrossRef]

13. Tzou, D.Y. A Unified Field Approach for Heat Conduction from Macro- to Micro-Scales. *J. Heat Transf.* **1995**, *117*, 8–16. [CrossRef]

14. Grmela, M.; Öttinger, H.C. Dynamics and thermodynamics of complex fluids. I. Development of a general formalism. *Phys. Rev. E* **1997**, *56*, 6620–6632. [CrossRef]

15. Öttinger, H.C.; Grmela, M. Dynamics and thermodynamics of complex fluids. II. Illustrations of a general formalism. *Phys. Rev. E* **1997**, *56*, 6633–6655. [CrossRef]

16. Pavelka, M.; Klika, V.; Grmela, M. *Multiscale Thermo-Dynamics Introduction to GENERIC*; de Gruyter: Berlin, Germany, 2018. [CrossRef]

17. Anufriev, R.; Gluchko, S.; Volz, S.; Nomura, M. Quasi-Ballistic Heat Conduction due to Lévy Phonon Flights in Silicon Nanowires. *ACS Nano* **2018**. [CrossRef] [PubMed]

18. Kröger, M.; Hütter, M. Automated symbolic calculations in nonequilibrium thermodynamics. *Comput. Phys. Commun.* **2010**, *181*, 2149–2157. [CrossRef]

19. Zhukovsky, K. Exact Negative Solutions for Guyer–Krumhansl Type Equation and the Maximum Principle Violation. *Entropy* **2017**, *19*, 440. [CrossRef]

20. Van, P.; Berezovski, A.; Fülöp, T.; Gróf, G.; Kovács, R.; Lovas, Á.; Verhás, J. Guyer-Krumhansl-type heat conduction at room temperature. *EPL (Europhys. Lett.)* **2017**, *118*. [CrossRef]

21. Pop, E.; Sinha, S.; Goodson, K.E. Heat Generation and Transport in Nanometer-Scale Transistors. *Proc. IEEE* **2006**, *94*, 1587–1601. [CrossRef]

22. Podlubny, I. Geometric and Physical Interpretation of Fractional Integration and Fractional Differentiation. *arXiv* **2001**, arXiv:math/0110241v1.

23. Li, C.P.; Deng, W. Remarks on fractional derivatives. *Appl. Math. Comput.* **2007**, *187*. [CrossRef]

24. Li, C.P.; Qian, D.L.; Chen, Y.Q. On Riemann-Liouville and Caputo Derivatives. *Discret. Dyn. Nat. Soc.* **2011**, *2011*, 562494. [CrossRef]

25. Li, C.P.; Zhao, Z.G. Introduction to fractional integrability and differentiability. *Eur. Phys. J. Spec. Top.* **2011**, *193*, 5–26. [CrossRef]

26. Cattaneo, M.C. A form of heat conduction equation which eliminates the paradox of instantaneous propagation. *C.R. Acad. Sci. I Math.* **1958**, *247*, 431–433.

27. Cattaneo, C. Sur une forme de l'equation de la chaleur eliminant le paradoxe d'une propagation instantanee (in French). *Comptes Rendus de l'Académie des Sciences* **1958**, *247*, 431–433.

28. Vernotte, P. Les paradoxes de la theorie continue de l'equation de la chaleur (in French). *C. R. Acad. Sci.* **1958**, *246*, 3154–3155.

29. Vermeersch, B.; De Mey, G. Non-Fourier thermal conduction in nano-scaled electronic structures. *Analog Integr. Circ. Signal Process.* **2008**, *55*, 197–204. [CrossRef]

30. Kovács, R.; Ván, P. Generalized heat conduction in heat pulse experiments. *Int. J. Heat Mass Transf.* **2015**, *83*, 613–620. [CrossRef]

31. Raszkowski, T.; Samson, A.; Zubert, M. Temperature Distribution Changes Analysis Based on Grünwald-Letnikov Space Derivative. *Bull. Soc. Sci. Lettres Łódź* **2018**, *LXVIII*, 141–152.

32. Sobczak, A.; Topilko, J.; Zajac, P.; Pietrzak, P.; Janicki, M. Compact Thermal Modelling of Nanostructures Containing Thin Film Platinum Resistors. In Proceedings of the 21st International Conference on Thermal, Mechanical and Multi-Physics Simulation and Experiments in Microelectronics and Microsystems (EuroSimE), 6–27 July 2020; in press.

33. Janicki, M.; Topilko, J.; Sobczak, A.; Zajac, P.; Pietrzak, P.; Napieralski, A. Measurement and Simulation of Test Structures Dedicated to the Investigation of Heat Diffusion at Nanoscale. In Proceedings of the 20th International Conference on Thermal, Mechanical and Multi-Physics Simulation and Experiments in Microelectronics and Microsystems (EuroSimE), Hannover, Germany, 24–27 March 2019. [CrossRef]

34. Raszkowski, T.; Samson, A.; Zubert, M. Investigation of Heat Distribution using Non-integer Order Time Derivative. *Bull. Soc. Sci. Lettres Łódź Série Rech. Déformations* **2018**, *LXVIII*, 79–92.

35. Curtiss, C.F.; Hirschfelder, J.O. Integration of stiff equations. *Proc. Natl. Acad. Sci. USA* **1952**, *38*, 235–243. [CrossRef]

36. Ascher, U.M.; Petzold, L.R. *Computer Methods for Ordinary Differential-Algebraic Equations*; SIAM: Philadelphia, PA, USA, 1998.

37. Süli, E.; Mayers, D. *An Introduction to Numerical Analysis*; Cambridge University Press: Cambridge, UK, 2003.

38. Auzinger, W.; Herfort, W.N. A uniform quantitative stiff stability estimate for BDF schemes. *Opusc. Math.* **2006**, *26*, 203–227.

39. Raszkowski, T.; Samson, A.; Zubert, M. Dual-Phase-Lag Model Order Reduction Using Krylov Subspace Method for 2-Dimensional Structures. *Bull. Soc. Sci. Lettres Łódź Série Rech. Déformations* **2018**, *LXVIII*, 55–68.

40. Raszkowski, T.; Samson, A.; Zubert, M.; Janicki, M.; Napieralski, A. The numerical analysis of heat transfer at nanoscale using full and reduced DPL models. In Proceedings of the International Conference on Thermal, Mechanical and Multi-Physics Simulation and Experiments in Microelectronics and Microsystems (EuroSimE), Dresden, Germany, 3–5 April 2017.

41. Bahrami, M.; Yanavovich, M.M.; Culham, J.R.; Thermophys, J. Thermal joint resistance of conforming rough surfaces with gas-filled gaps. *AIAA J. Thermophys. Heat Transf.* **2004**, *18*, 318–325. [CrossRef]

42. Bahrami, M.; Culham, J.R.; Yanavovich, M.M.; Schneider, G.E. Review of thermal joint resistance models for non-conforming rough surfaces. *AMSE J. Appl. Mech. Rev.* **2006**, *59*, 1–12. [CrossRef]

43. Volokitin, A.I.; Persson, B.N.J. Radiative heat transfer between nanostructures. *Phys. Rev. B* **2001**, *63*, 205404. [CrossRef]

44. Volokitin, A.I.; Persson, B.N.J. Resonant photon tunneling enhancement of the radiative heat transfer. *Phys. Rev. B* **2004**, *69*, 045417. [CrossRef]

45. Raszkowski, T. Numerical Modelling of Thermal Phenomena in Nanometric Semiconductor Structures. Ph.D. Thesis, Lodz University of Technology, Lodz, Poland, 2019. (In Polish)

46. Incropera, F.P.; DeWitt, D.P.; Bergman, T.L.; Lavine, A.S. *Fundamentals of Heat and Mass Transfer*, 6th ed.; John Wiley & Sons Inc.: Hoboken, NJ, USA, 2007.

47. Górecki, K.; Górecki, P. Compact electrothermal model of laboratory made GaN Schottky diodes. *Microelectron. Int.* **2020**, Paper version in press 3/2020. [CrossRef]

MDPI

St. Alban-Anlage 66

4052 Basel

Switzerland

Tel. +41 61 683 77 34

Fax +41 61 302 89 18

www.mdpi.com

Energies Editorial Office

E-mail: energies@mdpi.com

www.mdpi.com/journal/energies